Big Pharma, Dirty Lies, Busy Bees
and Eco Activists

Other books in the Stories from the Veld series by David Bristow

The Game Ranger, the Knife, the Lion and the Sheep (2018)
Of Hominins, Hunter Gatherers and Heroes (2019)

Big Pharma, Dirty Lies, Busy Bees and Eco Activists

Environmental Stories from South Africa

STORIES FROM THE VELD III

David Bristow

First Published by Jacana Media (Pty) Ltd in 2020

10 Orange Street
Sunnyside
Auckland Park 2092
South Africa
+2711 628 3200
www.jacana.co.za

© David Bristow, 2020

All rights reserved.

ISBN 978-1-4314-0900-6

Also available as an ebook

Cover design by publicide
Editing by Joey Kok
Printed and bound by ABC Press, Cape Town
Set in Minion Pro 10.8/15pt
Job no. 003726
See a complete list of Jacana titles at www.jacana.co.za

This book is dedicated to

Don, Mr Fix-It nonpareil, first class mate
and a good captain in a storm

and

Josephine, whose heart is big enough for
all the world and its failings,
including me and mine

Contents

Introduction
An Environmental Guide for Beginners – Including Fools, Drunkards and Assholes 1

1. The Man Who Invented Nature
Teach Your Children Well 5

2. Big Pharma and Dirty Little Lies
The Placebo Effect, Fools and Their Money 17

3. Big Business and the Dirty Business of Denialism
How Oil is Good for You, Why Luckies Prevent Throat Scratch and One Stubborn SOB 33

4. Rhinos Black and White
And Mambas Mostly Black, and Pack Hounds Mostly from Hell 41

5. Food for Thought I
Sugar and Spice and All Things Nice, Fatty and Salty and Many Things Nasty 49

6. Food for Thought II
The Broken Chain 59

7. The Real Costs of Power
Execrable Coal, Fricking Fracking, Super Solar or Wonderful Wind 69

8. FFS – The Dirty Little Company That Could
And the Green Bogeyman Who Couldn't 83

9. From GMO to OMG!
How Farming Evolved from Village Venture to Terminator Technology 93

10. Farmers from A to Z
Small Farmers, Good Farmers and Bad Farmers 105

11. Busy Bees
Dying Bees and Busy-as-Bees Honey Embezzlers 117

12. Rivers Blue and Brown
A River Used to Flow Through It 129

13. Plastic O Plastic
Macro, Micro and Now Nano – It's Time to Clean Up Our Act 141

14. The A, B and C of Climate Change
Anguish for the Atmosphere, Butchery of the Biosphere, Crying for the Cryosphere 151

15. Eco Activists
The Givers and the Takers, Superkids and Groan-Ups 161

16. Soil and Trees
Blood of the Earth and Breath of Heaven 169

17. Geological and Environmental Health
Why Where You Were Born and How You Were Born Can Really Matter 175

18. Green Living – and Dying
It's a Fine Art 187

19. Going Unplugged
Goodbye, Farewell, Totsiens, Hamba Kahle and Sala Kakhule – We're Headed Down the Road and Off the Grid 199

20. If Men Could Fly
What I Could Not Say 211

Postscript
A Handful of Stones 219

About the Author 223

Introduction

An Environmental Guide for Beginners – Including Fools, Drunkards and Assholes

A PHOTO DOING THE ROUNDS on social media some years ago shows a lad wearing only flip-flops, shorts and flamboyant sunglasses and holding a sign that reads: more trees, less assholes.

The lad in this photo – maybe 15 or 16, with his too-big, super cool sunglasses – is, I think, priceless. That sign he's holding says it all. I hope he is thriving and, who knows, is the leader of an environmental movement somewhere.

This book discusses the requisite subjects, including trees and climate change, rhino poaching and pollution. However, it also delves into less obvious subjects, such as some of the insidious ways food is produced, how we are lied to by big business almost every way we turn, and what happens after you flush the loo or put out your garbage bin. (Hopefully you wash your hands, for starters.)

Among the many wisdoms imparted by my mother, a very funny and smart woman of proud Irish provenance, was to never argue with fools or drunkards. Along with that sage advice, I have come to include conspiracy theorists, creationists and climate-change denialists who have learnt all they know from TV and the social media. This book is not for them: there is just no common ground there.

On the subject whether to vax or not to vax, it's more complicated. What it boils down to is understanding that there is good science and there is bad science. A lot of bad science has been used to bolster the anti-vaccinating information-stream, and that has led to a scepticism of the good science. Bear with me, and I'll strive to unpick it for you, like the difference between ethyl and methyl mercury, one of the anti-vaxxers' primary concerns.

It's a dirty business, this environmental muckraking and nit-picking, but someone has to do it. I hope my voice in the wilderness helps spotlight some of the wounds we have inflicted on Mother Nature.

David Bristow
Cape Town, October 2020

The Man Who Invented Nature
Teach Your Children Well

By the time Alexander Von Humboldt died, at approximately 2.30 pm on 6 May 1859, virtually every literate person in the world (the Western World at any rate) had been following his slow demise for months, then weeks, then days, and finally hours, as his body and spirit waned. His passing was mourned by tens of thousands of people in all the major European cities from London to Moscow, and several in America, both North and South. It was commemorated even in Cape Town.

He was the most famous person in the world, with more geographical features on Earth named after him than anyone else.[1] And yet, ask anyone today what they know of him, and they'll likely answer: "umm, the Humboldt Current?" He was the father of ecology – in fact, he invented the word. In a bigger sense, he could be thought as of the inventor of nature as we know it.

Every now and again, the slow evolution of human knowledge takes a quantum leap across the bridge of the unknown: when Galileo Galilei aimed the telescope, a new Dutch invention, at the night sky and saw craters on the Moon as well as moons around Jupiter; when an apple fell on the head of Isaac Newton and he thought, "Ah, gravity!"; when Charles Darwin looked at the tortoises and finches

---

[1] Certainly, a lot more than me. Although, if ever you find yourself in the dinky suburb of Duxberry, Sandton, David Avenue – there is my cadastral legacy. Our town planner father named the streets after the children who played there when it was still veld: Vincent, Glenn, Mervyn, Larry, Wendy and Anne, God bless you all, wherever you are now.

on the Galapagos Islands and thought, "Aha, evolution!"; and when Albert Einstein perceived the multi-dimensional curve of space-time.

Another, though less-known, moment was when Humboldt, with his life-long French friend Aimé Bonpland, stood at 19 413 feet altitude on the icy slopes of Mount Chimborazo in Ecuador (he measured everything he could), and the German had a revelation: everything in nature and indeed the universe was connected. This was not some ayahuasca-induced hallucination, but rather one of astute observation combined with brilliant insight.

Today we take for granted the principles of ecology, that all things really are connected. Pull on one thread of the universe, and you find yourself tugging at the entire fabric. This is what the explorer and naturalist perceived, standing in thin leather slippers, higher than any human had yet ascended. On his return to Germany three years later, he published this vision, along with detailed annotations, in a three-by-two-foot diagram titled *Naturgemälde* ("painting of nature"). It was published in his book *Essay on the Geography of Plants*.

That was but one of the titles in his 34-volume *Voyage to the Equinoctial Regions of the New Continent*. Over succeeding years, he wrote copiously, almost frenziedly – he was frenetic, witty, arrogant, garrulous and a compulsive worker. His books sold like none ever before or since. Among his more influential works were his political essays on Cuba and Spanish America.

He noted the inhumanity of slavery and how the monocultural colonial agricultural system was leading to widespread environmental degradation. Humboldt observed, researched, measured and documented what he saw of deforestation and the draining of wetlands. These works inspired Simón Bolívar to lead multiple revolutions against Spanish rule on the continent (he and Humboldt were one-time close friends).

But it was his life's final work that was – by modern assessment, at any rate – his greatest. This is *Cosmos: A Sketch of the Physical Description of the Universe*, and again *cosmos* was a term he invented. It took him 20 years to write the five-volume publication, which is hardly surprising given its scope: it is a virtual journey through Everything (stand aside and bow, Bill Bryson).

They begin with a discussion of the external world, starting with plants, volcanoes and fossils, followed by the extra-terrestrial one of planets, stars and nebulae. Then comes an internal voyage that meanders through the psyche of humankind from the Ancient Greeks to modern times: art, music, poetry, philosophy – everything is connected.

Hardly any aspect of life in the European realm was not greatly influenced by these works. Just about every naturalist of the time followed, in one way or another, in Humboldt's footsteps. Charles Darwin idolised him and carried his entire volume of *Voyages* aboard the *Beagle* (*Cosmos* had not yet been penned).

Without a Humboldt there could not have been a Darwin as we know him. The entire Romantic movement was inspired by the great German naturalist and explorer. Even the Art Nouveau style of the early 20th century arose in a direct line from him, through the illustrations of microscopic plankton by his acolyte, German marine biologist Ernst Haeckel.

And then Humboldt was forgotten, along with his all-encompassing view of nature, as specialised science ascended. But, lucky for us, that was not before his works had fired up and kick-started a new environmental awareness, particularly in the United States where his new ideas were appropriated like the wagons that, at that time, were rolling across the prairies.

Although John Muir, founder of the Sierra Club, is the most famous of Humboldt's American followers, even more influential in the mid-19th century were the writings of George Perkins Marsh (now also largely forgotten). Marsh was a ne'er do good autodidact (he spoke 20 languages) who revered Humboldt ("the greatest of the priesthood of nature"). Still, he could find no useful employment until he was dispatched to Istanbul as the US Minister Resident to the Ottoman Empire.

As a diplomat he was able to travel throughout the Middle East, often with his invalid wife Caroline. There he saw – and recorded in minute Humboldt style – how 10 millennia of agriculture had rendered the place barren. In his landmark work *Man and Nature*, he wrote that "assiduous husbandry of hundreds of generations had transformed this part of the Earth into an effete and worn out planet".[2]

But even back in his own, relatively inviolate country, Marsh saw how overfishing and damming were causing fish populations to plummet, how cash-crop agriculture was poisoning the land and the water, and how railway expansion and industry were eating up the forests and polluting the air. And, like Humboldt before him, he observed, and he measured everything he could. Way back in the 1860s and '70s, what he saw of stock farming convinced him that a vegetarian diet was less destructive environmentally than a meat-based one.

Also, very much in the vein of Humboldt, Marsh wrote "urgently, even rushed", because he felt time was running out to reverse the destruction of the natural world he saw in the places he visited (at the time in Italy's Po Valley).

From the lessons of the master, Marsh could see how, when milliners in Paris invented silk hats, and fur hats became unfashionable, it had a knock-on effect on decimated beaver populations in Canada. When farmers killed birds in large numbers (like when they exterminated the carrier pigeons) in order to protect their crops, they were then forced to use poisons to kill the insects and rodents that swarmed over their fields.

He noted that most of the killing of wild cattle and antelope for their hides, ostriches for their feathers, whales for their oil and elephants for their tusks was for the vanity of humankind – rather than necessity. Hardly surprisingly, therefore, his vision of the future of the planet was highly pessimistic.

He foresaw that Earth would be reduced to a barren planet, hammered by climactic extremes, and also possibly the extinction of the human species. Not enough people listened, or believed him then, more than 150 years ago, and just look at the pickle we are in now.

Yet that assessment is not altogether fair. Like Humboldt's, Marsh's book was in fact hugely influential in America at the time. It helped John Muir consolidate his wilderness vision and politicians of the day to put some brakes on the runaway depletion of natural resources, at

---

[2] *Man and Nature; or, Physical Geography as Modified by Human Action* by George P. Marsh (Sampson Low, Son and Marston, London, 1846)

least in the United States. It even encouraged foresters in this country to begin to conserve what remained of the limited natural forests of the southern Cape.

The works of both Humboldt and Marsh were central in influencing other great conservationists including Ralph Waldo Emerson and his neophyte Henry David Thoreau, the great nature poet Walt Whitman, bird authority and artist John Audubon, and then cascaded to the next generation through the likes of Aldo Leopold, Rachel Carson, EO Wilson and the maverick nonconformist Edward Abbey. So perhaps not all, or even very much, was really lost.

I CANNOT RECALL WHEN IT WAS I was baptised in the green waters of tree hugging and ant worship. I know it had something to do with the small stretch of veld where I grew up, running barefoot when lessons for the day were done. Small as it was, that limited patch of nature was still covered with wild grasses; small antelope took flight at our approach, and kestrels hunted overhead. The river that ran through the centre of it was really a stream (one of the bubbling brooks of the Witwatersrand), but it might as well have been the Zambezi for the hold it had on us children.

We didn't have many conservation icons back in those days; South Africans were far too entangled in a political *wag-'n-bietjie*.[3] There was in this neck of the veld a widespread tradition of big game appreciation, but not so much all the fine threads of nature. Nevertheless, there were still a few crumbs leading down the game trail of earthly and cosmic worship.

Visits to the various national parks made bush lovers of many of us who could get there. But my memory of those parks is that they felt like an extension of the Christian nationalist state replete with brown uniforms, shorts and long socks and everything closed during lunch time. Luckily, for many of us *Jock of the Bushveld* was our Beatrix Potter: dogs, baboons and crocodiles in place of rabbits, squirrels and ducks.

In our corner of the Highveld we had James Clarke, a senior editor

---

[3] *Wag-'n-bietjie* means "wait a while", the name of several hook-thorned trees and bushes in southern Africa that require you to take a moment should you get caught by their rapacious, recurved barbs.

of the otherwise staid newspaper *The Star*, who initiated the first urban conservation awareness project through his weekly CARE – clean air, rivers and environment – column. Clarke had grown up in some industrial town in England where, he once said, he could cut his lawn with nail clippers. When he first came to South Africa, he was entranced by the seemingly endless wilderness, but in time became distressed how little effort was being expended to protect it.

It was through his eyes that we started noticing how the tannery upriver was dumping toxic waste into our stream. Property developers along the outer edge of Johannesburg were bulldozing the natural vegetation so the ladies of the houses of the north could plant pretty English country gardens, and their men could water and fertilise and mow the lawns in the barren cycle of suburban life.

JAN SMUTS HAD GROWN UP on a farm in the Western Cape. For all his good and bad points, when the Second Prime Minister of South Africa decamped to Pretoria, he invited nature in, allowing the wild veld to grow right up to his stoep in Irene. He loved the veld, and the mountains, and for many he was an inspiration.

Among the first people to see and appreciate the destruction being wrought on the veld was a young adventurer named George Mossop. Sometime around 1870, he had run away from school and home in the Colony of Natal to seek adventure among the Boers of the Transvaal. He was there to witness the last of the great herds of plains game that used to migrate across the open grasslands.

In his thrilling autobiography *Running the Gauntlet*,[4] Mossop recounts seeing hundreds of thousands of springbok, blesbok, black wildebeest and zebras, including several hundred quaggas, of which he noted: "They were taller and of a lighter colour, the hair longer and more shaggy, with longer necks and bodies, and not so shiny and sleek, as their brothers, the zebra of the bush veld ... They were easily shot and were destroyed for their skins. This small troop of quagga were in all probability the last to be seen in Africa."

One day an experienced Boer hunter asked, could he hear anything?

---

[4] *Running the Gauntlet: Some Reflections of Adventure* by George Mossop (GCButton, Pietermaritzburg, 1990)

Mossop listened hard until finally he could detect a low rumbling like distant thunder.

"That is a thunder we must avoid," said the more experienced man. "You have not yet seen game, but today you will see it," and the party moved to higher ground somewhere in what is now southern Mpumalanga.

Soon there came a cloud of dust moving rapidly from east to west on a front three kilometres across. It was one massive herd of black wildebeest moving at a canter, some 60 kilometres long. It took nearly all day to pass. By that time, the Free State had been shot out, and after this there were none left anywhere.

Something else Mossop noticed was widespread environmental change fashioned by human hands. The area around Ermelo-Bethal-Chrissiesmeer-Wakkerstroom (now our country's coal-mine heartland) was covered by vast wetlands, extensive reed beds concealing tens of thousands of waterfowl – ducks, geese, teal, coots, cranes, herons and more.

Once the hunters were done, the farmers moved in. They burnt the reed beds, drained the marshes and planted thirsty trees, particularly gums, to lower the water tables. Early in the 20th century, Mossop wrote: "On these flats today stand many substantially built farmhouses, and on the marshy land herds of sheep and cattle graze the short grass. A stranger would not imagine that only a few years ago they were sheets of water."

For people born in the 20th century, our common infatuation with nature started with that grand old man of conservation, the Kruger National Park. For a majority of white South Africans, a trip to the Kruger became something of an annual pilgrimage; more often if you were so lucky. We could experience the wilderness, red in tooth and claw, even if we did not exactly appreciate how fortunate we were at the time, or how precious it was.

But it did not come easily. The battle to create the Kruger Park has been told in both previous title of this series (*The Game Ranger* and *Hominins, Hunter-Gatherers and Heroes*). Once he was given control of the iconic park in 1902, first warden James Stevenson-Hamilton didn't really have a clue what to do with it. His only orders had been to go down to the Lowveld and "cause as much trouble there as you can".

During the first two decades in charge, he and his fellow rangers had their hands more than full keeping people out of the park – subsistence farmers, ranchers, hunters and poachers (if they are not the same thing), miners, politicians and police officers with axes to grind and guns to blaze. Even they, the rangers, thought a large part of their job was to rid the park of all predators, or vermin, as just about everyone at the time thought of them.

Then one day, while he was sitting in the shade at Havemann's Hotel in Komatipoort, someone asked Stevenson-Hamilton when he was going to open his park for hunting. Suddenly the game ranger had an epiphany: "Never," he declared to the consternation of everyone else on the cool stoep. But then, why else was the government spending thousands of pounds a year just to keep animals safe, he pondered.

That was when it occurred to him: Why not save the place for nature's sake, for everyone's sake? He was aware of a nascent national parks movement in the United States and, probably more telling, by then the formidable Scotsman had fallen in love with the park he called his "Lowveld Cinderella".

"This set me thinking deeply," he wrote in his memoir.[5] What was the ultimate goal of a game reserve? "Would it be possible to turn the South African public from its present attitude towards wild animals, which was that of regarding them either as a convenient source of exploitation, or as an incubus hindering the progress of civilization?"

There was almost universal opposition to the warden's new ideas, but once again, the stroke of genius came over a glass at Havemann's. This time it was the proprietor who suggested that, if the warden proposed the name of Kruger for his new national park scheme, the heavily nationalistic Afrikaner cabinet under Prime Minister Louis Botha would find it hard to vote against, and he was proved right.

That was the first half of the 20th century, which was bisected into two mini centuries by world wars. Walking into the brave new world following the end of World War II were people like Ian Player, who had served in the tank corps in the American Army. (Many South

---

[5] *South African Eden: The Kruger National Park* by James Stevenson-Hamilton (Struik Publishers, Cape Town, 1993)

African soldiers were seconded to the US command for the invasion of Italy.)

Returning to his home in Johannesburg, Player found work underground in the mines. You can imagine how after each dynamite blast rang in his ears and clouds of toxic dust filled the subterranean tunnels, his thoughts drifted across the open veld and bush above, where the air was still clear and where one could dream.

In 1952 he applied for and got a job in what was then the Natal Parks Board (now Ezemvelo-KZN Wildlife). Here, Player found his own Cinderella, the Umfolozi Game Reserve. Clearly it was not all work in those days though, as he was a founding paddler of the Dusi Canoe Marathon (homemade wood and canvas boats in those days).

He rose up the ranks in the game department, and soon enough he was appointed warden of Umfolozi (now the Hluhluwe-iMfolozi Game Reserve). Prior to this, the country was set on an agenda of eliminating large game, both outside and within game reserves, in order to protect livestock from various maladies of the African bush. Player realised the southern white rhino population was down to about only 300 animals, most of them in his reserve, and facing extinction. He set in motion Operation Rhino – the human capture and relocation of the bellicose pachyderms – that almost certainly saved the species.

No doubt this episode got him thinking about larger issues, like the value of wilderness. In the late 1950s he lobbied successfully for the establishment of South Africa's first wilderness area, now the iSimangaliso Wetland Park World Heritage Site. Ideas about wilderness took him deeper into the human psyche, where he was greatly influenced by the works of Swiss psychoanalyst Carl Jung.

Player was one of the first people south of the equator to expound the idea that the health of the individual, as well as the national consciousness, was dependent on its relationship with nature and wilderness. With this in mind, he quit the parks' service and founded the Wilderness Leadership School, which greatly influenced a younger generation. From then to his life's end in 2014, he was the virtual leader of the wilderness movement in this country, as well as the patron of similar organisations on the rest of the continent, and also the United Kingdom and the United States.

Today, the veld of my youth is hardly recognisable, a great fruit pudding of human enterprise that is the largest metropolis in southern Africa, speckled with shopping malls and factories, mines and clutter housing as far as you can see. Imagine what George Mossop would have to say about it.

Under the cloud of this development, parts of the Highveld have among the worst air pollution in the world, with acid rain levels verging on vinegar – all those coal-fired power stations, with their multitude of problems, that don't work properly. Its rivers are running brown with sewerage and a white froth of plastic pollution. So many rivers in South Africa are contaminated beyond usable. Many others bleed red topsoil, the blood of the earth, into the Indian Ocean, first silting their estuaries before staining the adjacent sea.

In the mid-1980s, I took the royalties from my first book, *Mountains of Southern Africa* (it was, *quelle surprise*, rather a lot) and read for a master's degree in environmental sciences. Back then the scientific community was just starting to quantify – and to gasp at – the full extent of climate change and the disaster towards which we were catapulting our planet.

And now, some three decades later, in spite of overwhelming data and evidence, still not everyone gets it. Not everyone *wants* to get it. The "it" is that our lifestyles have become toxic to our home planet and the homes in which we live. We are taking a dump on our own front steps. In his award-winning documentary on the subject, former US Vice President Al Gore calls it an "inconvenient truth".

We would prefer not to be inconvenienced by the changes we would need to make to our lifestyle in order to turn back the tide of approaching doom. Is it really that bad? Yes, trust me, I'm an environmental scientist. (Then again, according to Donald Trump, you cannot trust climate scientists.)

But stay calm and pretty much carry on is one sage message. According to universal nature champion Sir David Attenborough, the changes we need to make are not, in fact, all that great. All we need to do is not eat that unnecessary last doughnut or pangolin, don't keep the tap running while brushing your teeth or throw away uneaten food along with the plastic wrappings. In three words: just *waste* less.

How hard could that be? For many people, too hard. It seems the more we have, the more we want, the more resources we use, and the more – much more – we waste. We are choking (much of the world is starving).

I have taken the title of this chapter from the Crosby, Stills & Nash song "Teach Your Children". It was part of the counter-cultural canon of the 1960s and '70s when young people realised their parents were messing up the environment – that their fathers' hell would indeed slowly go by.

They, their generation, my generation, has been brought up on wisdoms such as "silent spring", as set out by Rachel Carson. She showed how the DDT that was being sprayed wholesale across vast swathes of countryside and town – primarily to kill mosquitoes – was causing wide-scale die-off in fish and birds.

While not directly harmful to mammals, the chemical dichlorodiphenyl-trichloroethane, an organochlorine, accumulated up the food chain. The full effects were noted only when birds started dying en masse across the American landscape, most notably in wetlands – the veritable canaries in the environmental coal mine.

Crosby et al were hoping their generation would set things right. But clearly it didn't work out that way. The striving for an easy buck and increased convenience has proven to be a stronger incentive than cleaning your room. Why use a broom when God invented the leaf blower, because sweeping is just so hard? The jury is still out on disposable nappies (diapers for those across the big pond); they are among the most noxious of all modern conveniences.

As I write this, great swathes of the Amazon rainforest are ablaze, and the Brazilian government doesn't seem to give a cheeseburger. The president of the United States of America appears to be busy deconstructing that country's environmental regulations along with its wildlife and habitat preserves. People of my generation voted them into power. It boggles my mind, but there it is.

Then there are those, such as Space-X cadet Elon Musk (who, incidentally, hails from the Highveld) who are shooting for other planets. The problem as I see it is that places like the Moon and Mars don't have any rivers, or seas, or forests, let alone breathable oxygen. Surely it makes more sense to try to fix the one we're on!

But Musk is no blindfolded capitalist moneygrubber. He's a geek, no doubt about it, but he also has his eyes open to the ravages of the industrial world in which we now live. I suspect he sees the writing on the wall and is making ready for his own escape to Planet B.

I think it is already too late for my generation to fix things, and I have to admit our chances of reversing the devastating effects of climate change, pollution, habitat devastation, wildlife destruction and cascading health issues are not looking good. Take, for example, the simple plastic spoon.

WHO EVER THOUGHT IT WAS a great idea to drill into the ground to extract ancient oil deposits, send these to a refinery, turn them into plastic, in the process sending great plumes of carbon pollution into the atmosphere, then moulding that plastic into the shape of a spoon, or fork, shipping them to food stores all over the world at substantial carbon cost, only for these to be taken home, used once, and then thrown away… rather than washing an old fashioned metal spoon?

Insightful nature writer David Quammen likens the natural environment to a finely woven Persian carpet, all the threads attached to one another. If you pull one, you find it connected to all the others around it.[6] But now (primarily in the 20th century), we have taken a pair of scissors and cut the carpet into many small squares and separated each one with large tracts of "human footprint" – towns, cities, highways, farms.

What we are left with is not lots of small Persian carpets, but rather lots of small squares of pretty useless, fraying fabric. None of the threads is attached to the others any longer, and as the human population expands relentlessly, even the threads of the small pieces are being pulled apart.

Nature everywhere is in peril. While it is true that the human race could survive without nature, aloof in high-rise towers eating stuff made in laboratories, I would not wish to be part of that world any more than I'd like to live on Mars: no rivers, no forests, no seas, just more people living in oxygen bubbles. That sounds to me like Jean Paul Sartre's vision of hell. And that is why I am writing this book and, hopefully, why you are reading it.

---

[6] See *The Song of the Dodo: Island Biogeography in an Age of Extinction* by David Quammen (Pimlico/Random House, London, 1996)

I fear it will be up to our children to fix the machine we broke: How sad is that? But then I rejoice when I see the young ones shouting, "enough already!" Leading the charge is Saint Greta of Sweden. She's something of a geek herself, but maybe that is her strength: She has not allowed peer pressure, conventional wisdom or ridicule from people with powerful vested interests to quieten her.

Starting with a simple schoolyard protest against her elders for not doing enough about climate change, she has ignited a firecracker of youth protest around the world. She's dead right, and I pray that she and all those who follow her example have the courage, the strength and the perseverance to see it through.

We need more whistle-blowers, not leaf blowers.

FURTHER READING:

Abbey, Edward, *Desert Solitaire: A Season in the Wilderness* (Ballantine Books, New York, 1968)

Carson, Rachel, *Silent Spring* (Houghton Mifflin, Boston, 1962)

Dillard, Annie, *A Pilgrim at Tinker Creek* (Picador, London, 1974)

Leopold, Aldo, *A Sand County Almanac: And Sketches Here and There* (Oxford University Press, Oxford, 1989)

Muir, John, *A Thousand-Mile Walk to the Gulf* (Mariner, New York, 1998)

Quammen, David, *The Song of the Dodo: Island Biogeography in an Age of Extinction* (Pimlico/Random House, London, 1996)

Wulf, Andrea, *The Invention of Nature: The Adventures of Alexander von Humboldt The Lost Hero of Science* (John Murray, London, 2015)

Big Pharma and Dirty Little Lies

The Placebo Effect, Fools and Their Money

YOU'D PROBABLY EXPECT THIS BOOK to jump straight into the mucky business of pollution, polar caps melting and destruction of rainforests, but it doesn't. We'll get to all that dirty stuff in good time. But, to be fair, our environmental predicament did not begin with people wanting to do us harm.

When James Watt and Matthew Boulton developed a coal-fired steam engine, between 1763 and 1775, it was not with the intention of polluting the world's then pristine atmosphere. The same could be said about the 1970s "green revolution" that lead to the all-encompassing use of deadly herbicides and insecticides, which continue to poison us and the bees we so admire.

But it has certainly become a toxic relationship, that between us and big business – tobacco, oil, genetically modified (GM) farming, mass food production and the like. Like all harmful relationships, the environmental ones seem to start with lies: lies we are told when something is found to be unhealthy, like when your partner lies to you about the smell of booze on their breath, or a secret little bank withdrawal.

How about those old ads that told us smoking was good for us? Or that we had to take dietary supplements or risk our children being born with all kinds of mental problems? Or that if we ate a little capsule filled with basically sawdust, we'd get thin? At first these seemed like little lies, but once lying begins, there is no returning to good old times.

In a bigger picture, it's about how lies influence the greater body politic, which in turn knocks on to the natural environment. The ecological dictum, that everything is connected, is the underlying theme of this book. It's also about how (some) industries, (some) public relations companies and (some) advertising agencies collaborate to concoct and dispense these lies.

We may know some of the people telling these lies. Their children may go to school with our children, and they may spend their working days finding increasingly devious ways to bombard us with false and misleading information. The big question here is, why? Spoiler alert: Follow the money.

Let's admit it, all petrols are, like all mass-produced lager beers, distinguishable only at forensics level. It is the job of marketers and advertisers to lead us to want to buy one brand over another. At the same time, at least in the case of the petrol producers, there is a virtual arms race to make ever more efficient fuels.

I remember in the 1970s there was a product called Caltex with boron. I used to joke with the petrol station attendants and ask them to "fill up with Caltex with moron". I'm sure they didn't get the joke, although some of my mates swore it improved the performance of their 50 cc motorbikes. I'm pretty sure not many of them really knew what boron was, other than it was $B_{[5]}$ on the periodic table, or what it did in their petrol. It was a "deposit control additive" aimed to minimise the build-up of carbonaceous and inorganic lead deposits in the engine's combustion chambers.[1]

Each fuel company had its own additives to tackle this build-up and other engine performance issues. I suspected then, much as I do now, that "boron" was meant to sound scientific, mysterious and appealing. You could also choose Shell V-Power (nitrogen enriched), Engen Primax, BP Ultimate, Total Excellium or Sasol Turbo (with amaglug-glug, I think – the actual formulas were closely guarded secrets).

What would have been much more useful was if they had told us all their petrols contained lead. But that probably would have gone down with the driving public like ... a lead balloon. It's been known for

---

[1] https://pubs.acs.org/doi/pdf/10.1021/ie50504a053 (accessed 20 September 2020)

many centuries that lead is extremely toxic. Early in the 20th century, chemists working for General Motors proposed adding lead to their "gasoline", claiming it would greatly improve the performance of their engines.

They were dead right, and it proved to be a game changer for the internal combustion engine, allowing for significant increases in compression and heat, resulting in much more efficient and powerful engines. The motor industry as well as the petrolheads were thrilled, but not everyone was.

Some scientists were alarmed and urged the government to investigate the likely consequences. A leading expert in lead poisoning of the time, a Dr Alice Hamilton, sounded the alarm, saying it would be far too dangerous to even risk trying. "Where there is lead," she said, "some case of lead poisoning sooner or later develops, even under the strictest supervision."[2] Even within General Motors, the laboratory where the lead proponents worked was called "the loony gas building".

As I now head down my sixth decade, I am particularly aware of products aimed at the aging segment of the population, the one with retreating hairlines, advancing girths and aching joints. Many products containing MSM come to mind, never mind side effects including nausea, bloating and diarrhoea. MSM (methylsulfonylmethane) is a biologically active compound containing sulphur, which, many health websites tell us, boosts our immune systems and eases joint pain.

I went to several pharmacies in my area and asked the pharmacists about anti-aging medication. These are all people with degrees or diplomas in pharmacology. We inherently trust them due to those white coats. They each recommended various MSM-based products, yet the medical literature suggests that while MSM is safe, it "might" have some benefits to combat joint inflammation. It's likely no better for aching joints than smoking a joint. The only difference is one is far more pleasurable than the other.[3]

Ditto doctors. In his book *The Body*,[4] author Bill Bryson notes that

---

[2] www.bbc.com/news/business-40593353 (accessed 20 September 2020)
[3] en.wikipedia.org/wiki/Methylsulfonylmethane (accessed 27 May 2020)
[4] *The Body: A Guide for Occupants* by Bill Bryson (Penguin/Random House, London, 2019)

around 80 per cent of all antibiotics prescribed for humans are for complaints that cannot be cured by antibiotics. You'd think a doctor would know better, particularly with the demonic appearance of superbugs in our lives and hospitals. But again, no.

A doctor acquaintance admitted the reason he had closed his suburban general practitioner's rooms and instead gone into trauma surgery at a local hospital was because he had been losing customers steadily over the years. Why? Because all the parents wanted him to prescribe antibiotics for their little darlings whenever they tripped over a stick or sneezed. When he desisted, they took their business elsewhere; presumably to other doctors with fewer ethics and bigger cars.

And that's not even the worst of it: The majority of antibiotics dispensed in the world today are not even for human consumption, not directly anyway; they are destined for farm animals kept in such unnatural and debilitating conditions that they are continuously ill. We get all that medication for free in our farm products, thank you.

The use of methylsulfonylmethane was popularised in a book written by an American GP around two decades ago.[5] Author Stanley W Jacobs claims many people lack sulphur in their diets, and that he had proved MSM is a kind of miracle cure for all manner of aches and pains that resulted. He says the proof is in the tests he'd conducted on his own patients over the years.

But, media alert, when it came time to peer review his work, he had no data to show for all his tests. It appears that no sulphur deficiency has ever been described in any medical literature.[6] Nevertheless, products containing MSM continue to sell like koeksisters at a church bazar. (Probably the most effective commercial use of MSM is as a cutting agent in illegally manufactured methamphetamine.)[7]

The wide-scale acceptance of the benefits of MSM without what appears to be any supporting proof – or indeed in the face of facts to

---

[5] *The Miracle of MSM: The Natural Solution for Pain* by Stanley W Jacob (Penguin Putnam, New York, 2000)
[6] https://quackwatch.org/related/dsh/msm/ (accessed 20 September 2020)
[7] https://delphihealthgroup.com/drug-cutting/meth/ (accessed 20 September 2020)

the contrary – is strikingly similar to the issue of anti-vaxxing. Anti-vaxxing has its roots in Libertarian thinking, which holds that no government in any form should be able to tell me what I have to do or put in my body or not.

The things we are putting into our bodies, knowingly or otherwise, are frightening, and we have every right to be suspicious of vaccinations, for the sake of our children. However, we also have a responsibility to understand the minutiae of it, so we are very sure we are making the right decisions rather than following the loudest bleating herd. It boils down to knowing the difference between good science and bad science.[8]

For anyone who has not, it would be highly educational to see what widespread smallpox, polio, diphtheria or pertussis looks like before universally condemning the mandatory and universal vaccinating of children for dread diseases – as increasing numbers of people seem to be. While it does seem to be the case that, in a very small percentage of cases, vaccination affects neurological health in children, the consequences of not vaccinating against dread diseases are everywhere dire.

I have a memory of children at my primary school imprisoned in leg irons from having contracted poliomyelitis before we were all vaccinated. They were the lucky ones; they survived the disease. Decades later, I saw young adults in Malawi, crippled with polio, dragging themselves along the road or scooting on homemade skateboards while the developed world had forgotten all about this dreaded disease.

The main, most recent anti-vaxxing pushback began in Britain

---

[8] For anyone interested in looking deeper into this subject, here are some of the most pertinent references: *Bad Science* by Ben Goldacre (Fourth Estate, London, 2009); "Vaccines are not associated with autism: An evidence-based meta-analysis of case-control and cohort studies" by Luke E. Taylor, Amy L. Swerdfeger & Guy D. Eslick, *Vaccine,* June 2014; "Countering evidence denial and the promotion of pseudoscience in autism spectrum disorder" by Isabel M. Smith & Noni E. MacDonald, *Autism Research,* May 2017; "How autism myths came to fuel anti-vaccination movements: A timeline leading to the 2019 measles outbreak" by Eleanor Cummins, *Popular Science,* 1 February 2019 https://www.popsci.com/timeline-autism-myth-anti-vaccine/ (accessed 3 September 2020)

in opposition to the mandatory MMR (measles, mumps and rubella) vaccine for children. The anti-MMR movement based its beliefs on the work of one man, a doctor named Andrew Wakefield. In a February 1998 paper published in the esteemed medical journal *The Lancet*, Wakefield claimed he had found a connection between immunisation and autism. The media seized this, and it led to a wave of general vaccination hysteria.

Following widespread opposition from the medical profession, as well as counter lawsuits lodged by parents of children who had been vaccinated and had then showed signs of autism (some chronically), the General Medical Council investigated. It discovered that Wakefield had been paid by a legal firm to see if he could find evidence of a link, in order to use it in the various court cases.

Wakefield had cooked the books for financial reward. *The Lancet* retracted the paper, and the good doctor was disbarred from the profession. In February 2010, *Lancet*'s editor, Richard Horton, told *The Guardian* newspaper: "It was utterly clear, without any ambiguity at all, that the statements in the paper were utterly false."[9] Subsequently, English pharmacologist Dennis Flaherty called Wakefield's "the most damaging medical hoax of the last 100 years".[10]

But all this was hardly the beginning of the anti-vaxxing movement. Back in Victorian times, many people were horrified by the new-fangled idea of being infected with a deadly disease in order to be cured of it. For some people this horror has never gone away.

A more recent aspect of anti-vaxxing has been labelled "vaccine overload" – the belief that being giving multiple vaccines at once can weaken or even overwhelm a small child's immune system. "We weren't vaccinated against them," some parents today argue. "Many of us got mumps (ditto flu, measles, rubella and some others) and we all got over it, maybe even got stronger."[11]

---

[9] https://www.theguardian.com/society/2010/feb/02/lancet-retracts-mmr-paper (accessed 3 September 2020)

[10] "The vaccine-autism connection: a public health crisis caused by unethical medical practices and fraudulent science" by Dennis K Flaherty, *The Annals of Pharmacotherapy*, October 2011.

[11] "'Combined vaccines are like a sudden onslaught to the body's immune system': parental concerns about vaccine 'overload' and 'immune-vulnerability'" by

It is true we do not need to be vaccinated against all and every illness that comes along. It seems to be me (a veteran of many breaks and tears over the years) that getting a shot for tetanus every time you scratch yourself is about as bad as running to the doctor for antibiotics every time your nose runs.

In 1973, two Birmingham parents, Rosemary Fox and Renee Lennon, blamed their children's brain damage on the poliomyelitis vaccination. Media coverage of their plight found a groundswell of support, which led to the formation of the Association of Parents of Vaccine Damaged Children (APVDC). Over the next four decades, they raised awareness for their cause, and they also sought wide-scale recourse from the courts.[12]

With increased resistance to mandatory immunisation in Britain, an outbreak of pertussis (whooping cough) led to a nationwide study: every child there between two and three years admitted to hospitals for neurological illnesses was assessed to see if immunisation could be associated with increased risk. The finding was that it could – but that the risk was extremely low.

In light of, mainly, this study, a class action brought by the APVDC lost its case on the grounds of insufficient evidence linking vaccination to any harm done. But what about those who fall in the "extremely low" group? The emotional, social and monetary costs of caring for a severely autistic child can be crippling.

Over the pond in the United States, the anti-vaxxing crusade really got rolling after a 1982 TV documentary, *DTP: Vaccination Roulette*, and the 1991 publication of the book *A Shot in the Dark*.[13] As in the United Kingdom, anti-vaxxing advocacy groups were formed, and they sought recourse in the courts. The US cases were thrown out on the strength of contradicting evidence presented by the Academy of Paediatrics and the Centers for Disease Control and Prevention.

Shona Hilton, Mark Petticrew, Mark & Kate Hunt, *Vaccine*, May 2006.
[12] "A Disability Act? The Vaccine Damage Payments Act 1979 and the British Government's Response to the Pertussis Vaccine Scare" by Gareth Millward, *Oxford Journal of Social History of Medicine*, May 2017 https://www.ncbi.nlm.nih.gov/pmc/articles/PMC5410922/ (accessed 3 September 2020)
[13] *A Shot in the Dark* by Barbara Loe Fisher and Harris L Coulter (Avery/Penguin-Random House, New York, 1991)

In time, the campaign had been boosted by such medical luminaries as talk-show host Oprah Winfrey and actress and *Playboy* model Jenny McCarthy.

In February 2009, a Washington (DC) special court ruled that evidence in a group action presented by parents of stricken children failed to prove any link between autism and childhood vaccinations. By then, some 5 300 similar cases had been lodged in courts all across this litigious land. This ruling, as well as those in Britain, seemed to put the brakes on any further anti-vaxxing legal actions.[14] But then, suddenly, the tide seemed to turn…

In August 2020, while this chapter was being edited, news came from England that the mother of 18-year-old Robert Fletcher, a lad who had suffered brain damage when a toddler, had been awarded damages of £90,000. Since being vaccinated as a baby, Fletcher has been unable to talk, to walk or to feed himself. A panel of two medical experts and a judge concluded that, on a balance of probabilities, MMR vaccination was to blame.[15]

And then, while I was doing final corrections in September, news came that, after years of being denied, a US court had awarded the parents of 11-year-old Hannah Poling big time. When just 18 months old, because she had fallen behind in her vaccinations, she was given a five-in-one shot. It was found that that vaccination had exacerbated an underlying mitochondrial disease, which led to the onset of encephalopathy.

The court awarded the family US$1.5 million for vaccine-autism related damages as well as US$500,000 per year for her care. While the court found that the vaccination had not caused the "unknown mitochondrial disorder", it had aggravated the condition – and that was enough to prove cause and effect, legally.[16]

These two rulings will no doubt open the judicial floodgates. And

---

[14] http://edition.cnn.com/2009/HEALTH/02/11/autism.vaccines/ (accessed 3 September 2020)

[15] https://www.dailymail.co.uk/news/article-1307095/Family-win-18-year-fight-MMR-damage-son--90-000-payout-concerns-vaccine-surfaced.html (accessed 3 September 2020)

[16] https://www.cbsnews.com/news/family-to-receive-15m-plus-in-first-ever-vaccine-autism-court-award/ (accessed 3 September 2020)

perhaps this is fair justice for the small people, those who find they fall into the "extremely low" pit of children irreparably damaged after taking them for what they believed was a routine and indeed necessary minor medical intervention.

And yet, in 2019, the World Health Organization named "vaccine hesitancy" as one of the top 10 threats to global health.[17]

What emerges from this is that, while vaccinating for childhood diseases saves the lives of millions children a year, it also causes neurological complications in a small percentage. So, to vacc or not to vacc? Should we let nature take its course and kill off the weak? (We are told repeatedly that there are far too many of us and that it is modern medicine that has allowed population explosion to occur.)

If you were the minister of health, what would your ruling be here? To help you decide, here are some numbers, compiled by the US Centers for Disease Control and Prevention. Early in the 20th century, before the introduction of mandatory vaccines, there were 200,000 cases of diphtheria reported, whooping cough and measles more than 1 million each, and polio 20,000 cases. By the end of that century, numbers had fallen to just three for diphtheria, 1,500 each for measles and whooping cough, and seven for polio.[18]

Whether you ruled in favour of the masses or the few, there are still some concerns that have been raised by vaxxing sceptics that you would need to consider. For example, the occurrence of potentially poisonous metals, such as aluminium and mercury, found in vaccines. These two metals – much like selenium, zinc and copper – are, in small doses, essential to our good health, but are lethal in large doses. In the case of aluminium, most of us absorb up to 20 times the legal dosage allowed in vaccines from the environment, most of it courtesy of volcanoes. (See Chapter 17 for more on this.)

Then there is thimerosal, a form of mercury, which, we know, is extremely poisonous in concentrated amounts. It is there to prevent

---

[17] "What should we do about vaccine hesitancy?", by Fiona Godlee, *BMJ*, June 2019, https://www.bmj.com/content/365/bmj.l4044 (accessed 3 September 2020)
[18] https://www.forbes.com/sites/niallmccarthy/2020/05/14/how-vaccines-eradicated-common-diseases-in-the-us-infographic/#a7b1789721ff (accessed 3 September 2020)

the growth of harmful bacteria and fungi in stored vaccines. When it metabolises, it forms ethyl mercury, which is easily expelled from our bodies. Not so with methyl mercury, which could be extremely harmful to us. Appreciating the difference between the two is the difference between having passed or flunked chemistry – good science or bad.

As the green movement gains traction, and people become increasingly aware of what goes into the stuff that big business, big pharma, big whatever feeds us, we must strive to reduce the bad in order to increase the overall good. For example, mercury, in the form of thimerosal, good or bad, is no longer found in almost any childhood vaccines. It does, however, still occur in some applications, such as multi-dose flu shots. Best to take a few days off work and keep warm.

Both climate change denial and anti-vaxxing are cases where people are easily swayed by what *The New York Times* calls "Groupthink" parenting.[19] In our striving to sort the noxious chaff from the healthy wheat of modern living, it is often hard to know who or what to believe. It can be painful to admit the doctrine you based your faith on is mostly smoke and mirrors.

A final misunderstanding is that medical science is inextricably connected to the pharmaceutical industry. It's true there are some links. It's also true that some medical doctors are less pure than fresh snow, and that they sell their souls to Big Pharma. However, overall, medical science is an independent, loose and benevolent group of people pursuant of making humanity's journey through this world a painless one.

Parents have a right to vaccinate their children or not, just as they are free to bring them up with or without religion, as vegans or omnivores. The issue though is, what if a non-vaccinated child should infect others with a dread disease, at school for example? That family would almost certainly open itself to social sanction and even, possibly, expose itself to legal action?

THE GLOBAL MARKET FOR DIETARY supplements, vitamins and minerals is growing at an astonishing rate. According to a *Forbes*

---

[19] https://www.nytimes.com/2012/01/15/opinion/sunday/the-rise-of-the-new-groupthink.html (accessed 8 September 2020)

report, in 2018, nutritional supplements generated more than US$130 billion in sales worldwide. The market is predicted to nearly double by 2023. Today you can choose from a range of about 87,000 health supplements.

In South Africa in 2019, the "pills and snake oil" (aka supplements and vitamins) market was valued at R3.8 billion. From what I have read, the local medical profession is in step with overseas opinions: There is no evidence of their efficacy. Among the biggest sellers are immune system supplements. However, according to Cape Town endocrinologist Dr Jocelyn Hellig, our immune systems are extremely complex chemical factories on which no over-the-counter potion will have even the tiniest effect.

With regards to vitamins, from what I have read and come to understand, no one other than special people on special diets, such as vegans, people undergoing extreme treatments or people who follow a very unbalanced diet, should ever need to take extra. Vegans, for example, will not get vitamin B12, an essential animal protein, naturally in their diet, so they need to get it in the form of a supplement.

For the most part, it appears that the industry is built almost entirely with a narrative that pricks our two weakest spots – fear and vanity. You can't eat MSM tablets and expect your bones to get stronger any more than you can slap on face cream and get younger. Although there is a third, equally peculiar factor at play, and that is the placebo effect.

At some level we appreciate that our minds are the masters of our bodies. Think about how stress leads to all manner of health issues including, it is widely believed, the onset of various cancers. We are what we think we are, and can be just about whatever we think we want to be. All the motivational books tell us this.

It's an ancient maxim, yet whenever we see "mind over matter" applied in medicine, we tend to reject it. Call it placebo, it works in much the same way: There is no active medical ingredient, but if we think something is making us feel better, in many cases it will.

Through double-blind clinical trials,[20] we have come to learn that,

---

[20] In double-blind trials neither research subjects nor researchers generally know who is getting the placebo and who is not.

on average, a placebo is between 40 and 50 per cent as effective as drugs known to treat specific illnesses. They seem to work best on ailments such as depression, lower back pain and acute fatigue. Even domestic animals, particularly horses, have been shown to respond positively to placebos.

In the above-mentioned UK clinical trials, which focused on back and knee pain, some of it even debilitating, it was demonstrated that the "nocebo" effect was 25 per cent successful. In other words, even when patients had been told they were getting a placebo, they reported improvements. I am as susceptible as the next person to a placebo and have my own preferred snake oils. (It's my money. I'll spend it how I like.)

If you want to wind me up, just mention toothpaste or deodorant adverts. Why use just a bead of toothpaste when you can smear an entire caterpillar of it on your brush, or one spray of deodorant when you can spray your entire body and the entire room while you're at it? Don't you love it when someone gets into a lift and you choke on their scent?

One Axe spray deodorant ad in particular reminds me of the old joke about the young buck who asks the salesperson for a recommendation.

"Ball or aer'sol?" she asks.

"No, no, underarm," the millennial splutters.

The adverts show a ripped young man with his shirt off, spraying himself all over ball, aer'sol, underarm – the lot.

One day while I was watching an advertisement claiming some or other dishwasher used less water than hand washing the same load of dishes, the bell in my in-built crap detector alarm sounded. The bell in my in-built crap detector alarm sounded. An internet search revealed that the claim was indeed based on tests done at a suitably impressive institute in Munich. What they didn't say was that the hand washing was done with the tap running.

That is where most of the bad stuff is hidden, in the fine print. And that is why we all need to have our in-built, shockproof crap detector switched on at all times. I presented the information I found on water usage to the Advertising Standards Bureau, and they pulled the ads. Now I see the claim being made that only machines – and not hands – can get all the germs off your dirty dishes. There's no end in the ways they find to push our fear buttons.

Doing some hard time on the couch, researching TV adverts for this chapter, I was spurred out of the house to research a supplement that indicated it was "for cognitive support". I went down to the nearest pharmacy chain store and discovered that among its 20-odd ingredients are *Ginkgo biloba* leaf extract and… wait for it, boron.

As I further scoured the shelves, I found a "male performance booster" – with ginseng root extract and horny goat weed – that caught my attention. The active ingredient in horny goat weed, or barrenwort (*Epimedium sp.*) is icariin or, more technically, prenylated flavanol glycoside. In China it is known as *yin yang huo*, where it is sold along with rhino horn as an aphrodisiac. It seems it's about as effective as rhino horn, just much cheaper. In the only published medical trial I could find, icariin was found to have a mild antidepressant effect on mice.

There are so many others to comment on, like chromium picolinate, to help your blood release more energy: Who knew? There was another product with *Prosopis juliflora* var. *glandulosa* and mesquite for "blood sugar support".

I noted more shelf space was taken up by antioxidants than any other kind of supplement. It was touching to see parent shoppers piling their shopping baskets high with various vitamin bottles. If you read the medical and scientific literature, that which is not disseminated by the supplements industry, in my view there appear to be little to no benefit in these products. However, for many of us (call it 25 per cent or so) if the placebo effect works, it is worth the expense.

Something that might have appealed to me was one of numerous hair growth formulae. I did not have the heart or the inclination to tell the man putting it into his basket that one and only one thing can reverse balding, and that is follicle-stimulating hormone. Or that, if you were to build a human from a spares shop, that particular chemical would be the single most expensive part.[21] And believe me, I

---

[21] From "The Wine of Life" in *The Wine of Life, and Other Essays on Societies, Energy & Living Things* by Yale University molecular biophysicist Harold J

have done an enormous amount of research on this topic.

One of the most widely advertised beauty products on TV is shampoo. But back in the '70s, in the days of no-TV South Africa, we were told on radio the thing our hair needed most was Body on Tap with real beer. Apparently, what your hair needed most was a stiff drink. (They were right.) But beer was far too easy, and cheap, for the material age that followed.

Along with liquor, food and insurance, cosmetics has been described – by Emily Weiss, the founder of the US$1 billion beauty brand Glossier – as "recession proof". This is a US$500 billion industry, and they need to be inventing new stuff to keep the market captured; that's because we are nosey primates who are obsessed with the new and the shiny.

Nowadays you can buy shampoo with green tea, mushrooms, black truffles, honey, red ginseng and even animal placenta. Because, what your tired hair seemingly needs most these days is a bowl of molecular gastronomy. There are others with emu oil and even spermaceti, otherwise known as whale snot. In the past it was used for making candles and margarine.

In reality, shampoo is nothing more than good old-fashioned soap with some smellies added. But why buy simple, standard soap when you can buy extra-weak soap with all kinds of scents and fungi and embryos added? Shampoos are meant to remove three things from our hair: oil (sebum), skin flakes (desquamation) that come from our scalps, and dirt from the environment.

Shampoos consist mainly of liquid detergent and thickening agents, because watery shampoos do not sell well. People also like foam and think it makes for a better wash, even though it does not. What about those products that "add volume"? What they really do is coat your hair with some kind of goo.

Some added ingredients, like fish eggs, are just plain silly and aimed to part fools from their money. Others, though, can be downright harmful to both humans and the animals on which they are tested. This includes many fragrances, which come with a

---

Morowitz, (Abacus, London, 1979). Of the US$6 million Morowitz claims a person's parts would cost at the time the book was published, US$4.8 million was for the hair hormone.

long list of side effects. To preserve the beer, the fish eggs and the mushrooms, shampoos are often stabilised with sodium, sulphites or parabens such as 4-hydrobenzoic acid: just from the names you can tell they will not be kind to your skin.

Foaming and fast-drying agents are also bad for you – look out for DEA, TEA and the especially noxious foam booster cocamidopropyl betaine. If you see formaldehyde, remember it's primarily used for preserving dead things. Along with quaternium-15, it is a known carcinogen. Triclosan (banned in soaps but not shampoos) can cause hormonal disruptions.

The lists go on, but you get the drift. And you thought your shampoo manufacturer wanted to pamper you? That's not the case. Thankfully, nowadays there is an increasing array of genuine healthy soaps and shampoos available. When and wherever you find them, use them.

We all buy this stuff, often knowingly, because marketers are masters at playing on our dreams and our fears; fear that we are not doing enough for ourselves and our families, mainly our children, who need everything we can afford to give them a foot-up.

It's a crazy world and a complex one, and we have to be ever vigilant not be conned out of our cash at every turn and with every advertisement that accosts us. You are free to buy aqueous cream and family shampoo for around R100 together, or you can go big and spend thousands on much the same if it makes you happy. The results will be the same.

Big Business and the Dirty Business of Denialism

How Oil is Good for You, Why Luckies Prevent Throat Scratch and One Stubborn SOB

BACK IN THE LAND OF big oil with boron, companies have been helping us accelerate global warming for the past century. After scientific studies revealed direct links between automobile fuel and lead poisoning,[1] as well as between fossil fuels and climate change, denial went into hyper-drive. The industry, usually called oil and gas, has an annual value of US$1.7 trillion, which is about 4 per cent of global gross domestic product.

Among the first whistle-blowers against big oil was Charles Hamel in Alaska, who died aged 84 in 2015. Hamel started out working as a pro-oil political campaigner and then became an oil broker. However, the more he learnt about the sticky end of the business, the more sceptical he became.

Hamel became aware the Alyeska oil company was polluting Valdez Sound in Alaska with toxic sludge, and the air above it with equally toxic carbon gases. But no one believed him, given the fact that big oil was the major economic force in the region, and also given the huge donations it pumped into various social causes, the

---

[1] For her PhD in environmental sciences at University of Cape Town and subsequent work in the UCT Faculty of Medicine in the 1980s and '90s, Dr Yasmin von Shirnding found that children who attended schools alongside busy commuter routes had shockingly high levels of lead in their blood.

arts and political campaigns.

Hamel found disgruntled employees who secretly fed him the information he knew was lurking below the scum line. When Alyeska found out, they employed a security company to hound and threaten him.[2]

It was only after the oil tanker *Exxon Valdez* struck Prince William Sound's Bligh Reef west of Tatitlek in Alaska in 1989 and leaked 37,000 metric tonnes of crude oil into the bay that people in Alaska woke up to the fact of how negligent, irresponsible and even flagrantly criminal the oil companies had become.

The matter found its way to federal court in Washington DC, where presiding Judge Stanley Sporkin likened Alyeska's tactics to those of Nazi Germany. Anything to keep the oil flowing out and the money flowing in – including, it was implied, murder.

The oil industry has been seen as primarily responsible for fuelling the climate-change-denial lobby, with hundreds of millions of dollars allegedly being paid to smear good hard science, promote lunatic-fringe denialists, cover tracks and pay lobbyists in Washington to ensure the oil wells would keep on pumping.

Starting in the 1970s, ExxonMobil began funding organisations that would help delay the acceptance of global warming data.[3] The irony is that the industry's own global research did not do it any favours. For example, Shell analysts predicted "the disappearance of specific ecosystems or habitat destruction".[4] They warned that the negative environmental impacts of their employer would almost certainly be "the greatest in recorded history".

Among these impacts were devastating weather events, sea level rises and the desertification of large areas of the world. It made for terrifying reading, yet the oil giants still refused to disclose the true nature and likely impacts of their business. Luckily for the rest of us,

---

[2] https://www.washingtonpost.com/local/obituaries/charles-hamel-influential-oil-industry-whistleblower-dies-at-84/2015/04/29/e82df3ce-ee20-11e4-a55f-38924fca94f9_story.html (accessed 5 October 2020)

[3] www.theguardian.com/environment/2015/jul/15/exxon-mobil-gave-millions-climate-denying-lawmakers (accessed 20 September 2020)

[4] www.theguardian.com/environment/climate-consensus-97-per-cent/2018/sep/19/shell-and-exxons-secret-1980s-climate-change-warning

insiders leaked the findings to the media.

From my readings from the mid-1980s into the 2000s, Exxon pretty much led the way in climate change denial, lobbying strenuously against all and any legislation aimed at slowing global warming.[5] This included lobbying against the Kyoto Protocol, which aims to cap global carbon emissions. To this end, Exxon was a founding member and funder of the Global Climate Coalition (GCC),[6] comprising businesses opposed to the regulation of greenhouse (fossil fuel) pollution.

For more than a decade, the GCC bought adverts in the most influential newspapers in America including *The New York Times*, *The Washington Post* and *The Wall Street Journal*, casting doubt on the science of climate change. Between 1998 and 2004 it spent nearly US$16 million in spreading disinformation. Between the mid-2000s and 2015, it paid more than US$2 million to climate change deniers in the US government.

Sarcastically, the point was widely made that a plucky band of oil billionaires and politicians were the ones to expose almost all of the world's scientists, who were conspiring to create the environmental crisis "hoax". It was not, however, the oil companies that invented these obfuscatory tactics. They learnt everything they know from American tobacco companies.

It seems extremely funny to us now when we see old adverts that tell us things like, "Nine doctors out of 10 recommend Camel cigarettes." Satire from the pages of (the now sadly defunct) *Mad* magazine? Unfortunately, not. Copywriters in advertising agencies were doing what they do best.

The line above is straight out of a real advertisement for the RJ Reynolds tobacco company, founded in 1946 (the same year, co-incidentally, that *Mad* was first published), and seen in leading

---

[5] I have long been bemused by the mainly American lobbying system, where monied interests pay specialist professional lobbyists to influence politicians in favour of their business. The fabulously wealthy Charles and David Koch spent around US$130 million of their family's foundation money to directly fund close on 100 climate denialist people and organisations. But they are far from being alone, and you can add the names Anshutz, Bradley, Coors, DeVos, Dunn, Howard, Pope, Scaife, Searle and Seid among the really big spenders.

[6] https://documents.uow.edu.au/~sharonb/columns/probe11.html

magazines such as *National Geographic* and even the official journal of the American Medical Association. The "independent surveys" on doctors' smoking habits was conducted by William Esty Co, advertising agent to none other than RJ Reynolds.

Tobacco advertising in the early 20th century made even more amazing claims: that smoking boosted athletic performance, that it was good for pregnant women, that it prevented hangovers, it soothed sore throats (say *what*?) and that it was safe for babies – not smoking, but secondary smoke.

It seems funny now, but it was not really funny at all because the tobacco companies at the time knew exactly what they were doing, not just selling dangerous tobacco and nicotine, but adding other ingredients – including cyanide – to make their cigarettes ever more addictive. There has been much written and many documentaries made about cigarette companies and the lengths they went to to persuade us that cigarettes weren't harmful.

It's been reported that these companies paid people, first in the advertising industries to lie about it, and second, in the marketing business to blow a smokescreen over it and then, third, supposed experts and even scientists to create false narratives about smoking and so discredit the truth wherever and whenever it seeped out.

Can we believe the scientists when they say: "20,679 physicians say Luckies are less irritating; it's toasted. Your throat protection against irritation, against cough." This is also a real ad.

How about this for scientific muddying, poor grammar notwithstanding: "Chesterfield: new scientific evidence on effects of smoking. A medical specialist is making regular bi-monthly examinations of a group of people from various walks of life. Forty-five per cent of this group have smoked Chesterfield for an average of over ten years… After ten months, the medical specialist reports that he observed no adverse effects on the nose, throat and sinuses of the group from smoking Chesterfield."[7]

When medical evidence overwhelmingly revealed smoking as

---

[7] http://www.history.com/news/cigarette-ads-doctors-smoking-endorsement (accessed 20 September 2020)

a serious health hazard, tobacco companies started promoting the one characteristic it still could: smoking could help keep your weight down. Clearly targeting those trying to lose weight – including new mothers and sports players – smoking ads in the late 1960s and early '70s were full of allusions to slimness, slim models and slim cigarettes.

Then in the 1970s, cigarette ads were full of racing car drivers, cowboys, exotic adventures and outdoorsy people. There was the Marlboro Man and the Camel Man. They were very carefully designed to recruit an emerging market of smokers, namely preteens.

You have to wonder what the people who worked in the tobacco industry, the people who worked for its advertising companies, thought about it, and told their children and their friends they did for a living. As was the case in the oil industry, one of the first whistle-blowers was an insider, Jeffrey Wigand. His own children are said to have asked him if the cigarettes he made for a living really killed people.

That got him thinking. You might recall the movie about Wigand, "the man who knew too much". It was titled *The Insider*, with Wigand played by Russell Crowe. It also featured Robert de Niro, Christopher Plummer and Mike Moore as himself.

Wigand was a chemist and vice president for research and development at Brown & Williamson (B&W), a subsidiary of British American Tobacco (BAT), which became the world's largest tobacco company based on net sales in 2019. It is located in Louisville, Kentucky. BB&W was in turn owned by BAT, the third largest in the world, and the one known by insiders as "the tough guy" due to its aggressive litigation strategy, BAT also donates vast sums to various causes, museums and politicians.

Wigand became vocal about cigarette additives (such as ammonia, which increased the effects of nicotine) that made them even more addictive and carcinogenic. When he openly questioned B&W's additives policy, he was put on notice and, in 1993, fired – but not before being forced to sign a lifelong non-disclosure contract with B&W.

What followed was a long, circuitous and scary story. Wigand became a consultant for anti-smoking groups. BAT sent in its big guns, hiring an international investigative company to dig the dirt on him, portraying him in a 500-page document as a thief (he'd once stolen a bottle of booze), a wife beater (in divorce papers, his first wife claimed he'd hit her with a coat hanger), a drunk, a cheat and a plagiarist.

But worst was when they set the pit bull of New York PR onto him, a man named John Scanlon, whose job was to publicly discredit Wigand and, according to the whistle-blower's brother, the PR man branded him a "habitual liar" and a "bad, bad guy".[8]

Then the death threats started: "We want you to know that we have not forgotten you or your little brats," and "If you think we are going to let you ruin our lives, you are in for a big surprise! You cannot keep the bodyguards forever, asshole." This was from people working in public relations.[9]

It might sound all too familiar to South Africans, who will recall the Bell Pottinger scandal involving the British PR company that was hired by then-president Jacob Zuma's enablers, the Gupta brothers, to discredit anyone who stood in the way of their state-capture gravy train. It was BP that dreamt up a smokescreen narrative about a white monopoly capital conspiracy that was bent on discrediting the Zuma gang.[10]

Only after the crusading American TV news programme *60 Minutes* featured Wigand and his revelations did Washington get serious about tobacco. In 1994 Congressman Henry Waxman was holding public hearings on tobacco when Wigand, following them closely, realised the huge degree to which his old employers were lying under oath. "They lied with a straight face," he said on *60 Minutes*. "Sandefur [his former boss at B&W] was arrogant! And that really irked me."

BAT sued Wigand for US$60 million. The old chemist counter-sued. "Leave or else you'll find your kids hurt," orchestrated Scanlon. But by then the smoke was pouring out from under the signal blanket.

---

[8] https://www.vanityfair.com/magazine/1996/05/wigand199605 (accessed 4 September 2020)

[9] *Ibid*

[10] In January 2020, a documentary titled *Influence*, produced and directed by *Daily Maverick* staffers Richard Poplak and Diana Nellie, premiered at the Sundance film festival in Colorado. The full-length movie focused on Bell Pottinger and "serves as a terrifying reminder of the dangers that lurk within the post-truth era, in which masters of disinformation use new digital tools to wage ancient propaganda wars."

BAT tobacco's lie had been called.

In 2006, Washington District Judge Gladys Kessler issued a 1 600-page opinion that ruled the tobacco industry had engaged in a decades-long racketeering enterprise. That's akin to conducting organised crime. She ordered the various companies on trial to take a number of actions, such as ceasing to claim their products were not harmful, or less harmful than any others, and targeting youth. They were also slapped with massive fines and penalties.

But big tobacco was not going to just lie down and cough up. The companies appealed and fought back with vigorous advertising and PR campaigns. Eventually, the slow-turning wheels of justice had them ground down to ash, and in 1998 they reached a settlement of US$206 billion, to be paid over 25 years, with the attorneys general in 46 US states.

After years of not being able to find employment, in the aftermath of the smear campaign waged against him by B&W, Wigand's story does have a silver-wrapped ending. He found a job teaching at a high school in Louisville, and was voted the top teacher in the state of Kentucky in 1996.

Wigand's brother is recorded as saying: "He is one stubborn SOB; the tobacco people picked on the wrong guy."

Rhinos Black and White
And Mambas Mostly Black, and Pack Hounds Mostly from Hell

EARLY IN THE 20TH CENTURY, there were about 500,000 rhinos, mainly black but also a few thousand white, roaming the African savanna. By 1970, the number had dropped to an estimated 70,000, while today there are estimated to be fewer than 29,000 – about two-thirds of them in South Africa. It is hard to give more precise figures, since the two species are the targets of the most concerted war against wildlife ever conducted by human aggressors.

Annually, the trade in rhino horn is worth around R4 billion, which is why so many people are willing to risk their all in its pursuit. One horn has a street value of around R4 million. What is so bitterly sad is that it is made of basically the same stuff as our hair and nails – keratin – no more and no less, and of use only to a rhino.

The most amazing, alarming, upsetting and informative wildlife documentary I have ever seen delves deeply into the dark world of rhino poaching and the world trade in rhino horn. It is titled *Stroop: Journey into the Rhino Horn War*. In Afrikaans, the verb *stroop* means to poach.

The *Stroop* team follows the money from the killing fields in the Lowveld and Zululand, through the street markets of Jozi – where you can find rhino horn, leopard skin, and pangolin, vulture and lion parts all openly on sale – all the way to the street markets of Vietnam and finally to the ivory and horn "factories" deep in rural Laos.

In southeast Asia, they found horn, horn-grinding bowls and ivory trinkets "everywhere". One haunting sequence shows ivory –

hacked out of dead, bullet-ridden elephants, some while the animals are still alive – later being turned by Laotian workers into trinkets for mostly Chinese tourists.

From the documentary, one learns that after China delisted rhino horn from its official register of traditional medicines in the 1990s, crime syndicates created new markets. First was Taiwan, an eager receiver, until it was banned there. After Taiwan, efforts were moved to Vietnam, where an aging politician and war hero, claiming that horn powder had cured his cancer, became the poster boy for the rhino horn industry.

Rhino horn sits alongside tiger, lion and bear parts, pangolin scales and other bits and pieces of many other dead, once wild, creatures on the streets of the Vietnamese capital Hanoi. It is believed to cure anything from cancer and fever to a hangover. Vietnam has one of the world's fastest growing wealthy populations, and rhino horn powder has become the new cocaine at fashionable dinner parties. Seen as a status symbol, rhino horn and the paraphernalia needed for its consumption are all extremely expensive.

There have been so many reports about the impunity with which the smugglers or poachers seem to operate and it seems as if the political will to bring poachers and others in the rhino horn trade to book often seems to be lacking. The delay in the prosecution of alleged poaching kingpin Dumisani Gwala is a case in point. The fact that he and his accomplices face many charges related to rhino poaching, and that in spite of numerous court appearances, he has managed to evade the full force of the law, has frustrated rhino conservationists.

Following the legal paper trail is fascinating, and if the many reports in newspapers such as the *Sunday Times* and the *Independent* are to be believed, then the story takes us first to the Mtubatuba Magistrate's Court. Here, environmental activist Jamie Joseph alleges that Magistrate Deuteronomy Ngcobo "has a long history of letting off rhino poachers with a slap on the wrist or a small fine".[1] Her investigations appear to link Magistrate Ngcobo with the office of

---

[1] https://www.savingthewild.com/2016/01/rhino-poaching-kingpin-and-the-magistrate-that-keeps-him-out-of-jail/ (accessed 17 August 2020)

regional court president Eric Nzimande.

After many delays, in April 2019, Gwala finally had his day in court… Well, almost. His co-accused, Aubrey Dlamini, told the sitting magistrate he was not feeling well, and once again the case was postponed.

In December 2019, Nzimande was suspended from duty, pending investigations into alleged bribery and corruption.[2]

Some of our national politicians have been associated with the rhino horn trade. David Mahlobo, a former minister of energy and state security, is a man wallowing deep in this controversy. When caught on camera visiting the resort of Chinese gangster and third-tier rhino-poaching intermediary Guan Jiang Guang in Bela-Bela in 2016, the minister told his accusers he had merely gone there for a spa treatment.

In one of Gwala's court appearances, a former Hawks officer, Detective Warrant Officer Jean-Pierre van Zyl Roux explained that, as he saw it, the illegal rhino trade has four tiers.[3] First there are the hunters, their guides and carriers (guns, water, machetes and axes). The second tier includes the hunt organisers, transporters, traditional healers who "bless" the hunt and sometimes receive small amounts of illegal horn, and then the game rangers who turn money-blinded eyes, or supply inside information about rhino locations and anti-poaching deployments to the hunters on the ground.

Next comes the first order of intermediaries who move the horns through commercial and border hubs to the higher level of intermediaries.

When people implicated or complicit in rhino and elephant poaching include game rangers, police officers, magistrates, politicians – and even heads of state – what chance do the poor animals have, or those brave warriors determined to protect them? Guang was subsequently reported to have fled the country under cover – probably official cover – but we might never find out the full story.

---

[2] https://www.iol.co.za/sunday-tribune/news/kzn-regional-court-president-suspended-amid-bribery-allegations-19770027 (accessed 20 September 2020)
[3] https://www.timeslive.co.za/news/south-africa/2019-04-25-rhino-poaching-kingpin-finally-in-the-dock-after-20-delays/ (accessed 20 September 2020)

ONCE UPON A TIME, THERE was a wild African unicorn. It was called *Ceratotherium simum cottoni,* the northern white rhinoceros, but only two of them survive in captivity, protected night and day by men and women with modern weaponry.

There still is one kind of unicorn currently living in Africa, the southern white rhino (*Ceratotherium simum simum*), a most improbable survivor from the Pleistocene past. Unfortunately, if we continue at this rate, in a few decades' time, it too might go the way of the dodo, blue buck, tiger, the pangolin, the Ethiopian wolf, the African wild dog, the cheetah, the dinosaurs – and all the other unicorns.

For the past few million years, white rhinos roamed the grasslands of southern and east-central Africa, moving like deliberate two-tonne lawnmowers. With no natural predators they lived mostly in peace, but they bred fairly slowly. Then, in the mid-19th century, people with guns arrived on the scene, and by the mid-20th century, the southern race was extinct across most of its range. The northern race was doing only marginally better at that time, given their naturally languorous ways.

In the late 1950s, a game ranger in the Zululand region of South Africa named Ian Player (brother of the more famous golfer Gary), realised his park had the last viable population of maybe 200 southern white rhinos. However, breeding at best only about once every two years, their chances of long-term survival were looking bleak.

Working in what is now known as the Hluhluwe-iMfolozi Game Reserve, Player cobbled together a team with the intention of breeding and moving viable family groups to other game reserves in the region. Operation Rhino had to pioneer all the tricks of the "trade", of capturing and relocating what in the field are known as mega-herbivores. Those were fun times, and many a well-muscled man swallowed mouths-full of pride and dust.

The largest satellite herd went to the Kruger National Park. Today the Kruger safeguards – or tries to – a population of between 6,000 and 7,000 of the wide-mouthed horned beasts. It is hard to be precise, since every year over the past decade or more poachers have killed several hundred.

Currently, more than three-quarters of the entire remaining number of the sub-species (although, given that the northern sub-species is extinct in the wild, we might as well say species), which totals maybe 20,000 individuals, can be found in South Africa.

The story of Operation Rhino, bringing back the species from the toe-edge of extinction, was a nail-biting one, but it's not over yet. The current poaching is a veritable tsunami of human cruelty and greed. Since rhino poaching really hit hard in the Kruger Park, around 9,000 rhinos have been killed there: a third of the total world number.

One minor irony of this story is that the southern white race, once the most endangered of all rhinos, is now the most numerous of all.

Poaching and anti-poaching is an arms race. Poachers on the ground risk their lives for a resource that can fetch more than gold or heroin on the street. Not even zoo-held animals are safe anymore.

Further muddying the waters are horn-trade speculators, who have lobbied conservation agencies in some African countries, and are now doing the same on other continents, that trading horn on the open market is the best way to save rhino. There are many cracks in this classic economic model, not least the aforementioned fact that criminals control the market.

An average of around 1,000 rhinos have been poached each year in South Africa over the past decade (the peak in 2014/2015 was more than 1,300).[4] This would yield about 2 tonnes of horn. By my estimates, private breeders in this country have around 2,000 animals bred as you would cattle. Apart from any stockpiles they might have, they can supply about a quarter of that amount. When the largest of the breeders tried so sell 500 kilograms online he admitted there were "very few bidders". Why buy it at street value when you pay a poacher a pittance to get it for you, seems to be the wisdom here.

The breeders are known to ship animals illegally to China where, it is said, they will be protected like they were walking gold and the horn harvested. Meanwhile the animals are kept in concrete pens resembling pig sties.[5] Could this seemingly ugly scene actually

---

[4] www.savetherhino.org/rhino-info/poaching-stats/

[5] National Geographic, "Special Investigation: Inside the Deadly Rhino Horn Trade" by Bryan Christy, Oct 2016, www.nationalgeographic.com/magazine/2016/10/dark-world-of-the-rhino-horn-trade/ > "Meanwhile a booming illegal trade supplies mostly Vietnam and China, where rhino horn is often ground to a powder and ingested as a treatment for everything from

represent the future of the species?

Meanwhile, back in Africa, the daily battle against rhino poaching has continued. As anti-poaching units upped their game, with better trained and equipped personnel and higher tech solutions, killings came way down toward the end of the second decade of the 20th century. At the same time, poacher apprehension and arrests went up significantly (from four or five a year to around 150 in 2019).

It's gone to the dogs, quite literally. K9 is an anti-poaching unit operating in KwaZulu-Natal's Zululand district, the combined efforts of 18 state, community, public and private wildlife concerns. It's a quick-response unit using aircraft, dogs and horses. A large part of its work is community outreach and education in the region, highlighting the plight of the rhino, and the pangolin. South African National Parks has also seen a significant spike in poacher arrests with their own canine anti-poaching units. A South African Wildlife College initiative in Hoedspruit has also had some good results. The main problem with these canine units has been that they are only as fast as their weakest link – the human.

Initially Belgian shepherd or Mallinois were used in human-dog teams, but their success rate was not stellar. Investigations into dog-pack use led to Joe Braman, a trainer, rancher, law officer and breeder of Texas coonhounds, which are bred to hunt raccoons. Similar in looks to bloodhounds, the coonhounds are bred and trained to be much fiercer. Braman grew up training free-running packs of coonhounds and says: "It's all about intimidation. If one of these dog attacks you the first thing you're going to do is throw the gun and climb a tree."[6]

The Texan dog man was convinced his especially aggressive pack hounds could solve the Kruger's poaching problems,[7] but the park's chief dog trainer, Johan van Straaten, was not convinced. Killing poachers – which Braman's dogs were initially trained to do – was

---

cancer to sea snake bites and hangovers."

[6] https://www.nationalgeographic.com/animals/2019/09/texan-hounds-fight-rhino-poaching (accessed 17 August 2020)

[7] http://testing.environews.tv/world-news/091819-texas-pack-hounds-charge-to-the-rescue-for-rhinos-in-south-africa-nabbing-145-poachers-so-far/ (accessed 20 September 2020)

not the Kruger's objective. "They're [Braman's team] really hard on their dogs. They work with whips. Shouting at dogs – shocking them if they didn't do the right thing."[8]

A somewhat less aggressive approach notwithstanding, in their first year of operating in the park, the Texan hounds helped apprehend 145 poachers and seize 53 firearms, more than 10 times the number apprehended by the older canine units.

And if it's not going to the dogs, then it's going to the mambas. The all-women Black Mamba Anti-Poaching Unit patrol the Balule Nature Reserve, which forms part of the greater Kruger Park and is sandwiched between the Klaserie and Timbavati reserves.

Established in 2013, the Black Mamba team patrols the reserve boundary for poachers or their snares, as well as fence breaches. On accepting a World Tourism Conference award, Black Mamba founder Craig Spencer noted: "Coming from disadvantaged communities and breaking strong patriarchal tradition, these courageous women focus on eliminating illegal wildlife trade through conservation, education and the protection of wildlife."

Poachers are to be feared, but the Black Mamba say they are more afraid of what would happen to the animals if they were not there. They want to change not only the lives of the rhino but also their own, bettering their futures and changing the attitude of the communities around them.

"They say it's a man's job, but we are doing it," Felicia Mogahane, one of the Mambas, told *National Geographic*.[9] And none too soon. Maybe that explains why we have a Father Time, destroyer of all things, but a Mother Nature, nurturer of all life.

---

[8] https://www.nationalgeographic.com/animals/2019/09/texan-hounds-fight-rhino-poaching (accessed 17 August 2020)

[9] https://www.nationalgeographic.com/adventure/destinations/africa/south-africa/black-mambas-anti-poaching-wildlife-rhino-team/ (accessed 17 August 2020)

Food for Thought I

Sugar and Spice and All Things Nice, Fatty and Salty and Many Things Nasty

Warning: This chapter contains colourants, additives, starches, extracts, enzymes, added sugar and salt and might be contaminated with nuts and other ingredients that could be harmful to your health.

Have you ever wondered why supermarket peppers are always so shiny, why water and protein have been added to your favourite porkies, why the cherries in your cake stay fresh and firm forever, and how a can of cream can keep on a store shelf for months? Because they have all been augmented in some way, of course.

Increasingly over the past several decades, factory-made foods have become a staple in our diets. Even in rural South Africa, the old subsistence-food-production systems are giving way to cash economies, where most of the cash ends up in the local supermarket or fast-food outlet.

While we all appreciate that natural foods are best suited to giving us rich and sustaining lives, we have become addicted to more convenient but inferior "fast foods" that pack in sugar and salt. It's no mystery, more of a dark art. A conspiracy theory, if ever there was one, to make us buy and eat more and more of less and less.

Have you ever wondered why, for example, when all you need to make mayonnaise at home is oil and egg, with maybe a pinch of salt and a squeeze of lemon juice, most supermarket types have between

10 and 20 ingredients?[1] Unless you are a chemist, you will not know what all of them are. Even the egg on the label might turn out to be only a distant relative of anything that ever came out a hen's cloaca.

In her book *Swallow This*,[2] English food writer Joanna Blythman tells us that sales reps for food ingredient manufacturers carry heavy volumes of flavour libraries and veritable Pantone charts for colourants to manufacturing conventions. No processed food (meats, fish, shellfish, right through the supermarket shelves to chips) escapes their all-consuming inclusions of infusions, coatings, glazes, bastings or dustings.

You'd think a fruit-flavoured yoghurt for breakfast would generally be good for you, but again you're being misled. A 2019 British National Institute of Foods (BNIF) study[3] identified these yoghurts, particularly the low-fat ones (which by default are often extra high in sugar), as being among "super processed" foods linked to obesity, diabetes, cancer and heart attacks.

Besides yoghurt, the study's list of super processed foods includes fizzy and energy drinks, margarine, factory breads, chicken nuggets and hot dogs. Even granola bars, that go-to "health" snack for many sporty people, are blacklisted by the BNIF. The study shows that just about all processed foods (anything with added colour, flavour, oil, sugar or salt) actually causes us to eat more and gain weight.

It is repeatedly pointed out that all food consists of chemicals – as do we all; we're veritable walking, talking periodic tables. But what we don't want is too many molecules of things such as sodium ascorbate,

---

[1] Most mayonnaise contains vegetable oil (52%), canola seed, anti-oxidant TBHQ (E3119), water, sugar, stabiliser (E1422, E1450, E415, 412), pea protein, salt, acidity regulators (E330, E270), colourant (E160a, 270), preservatives (potassium sorbate E202, sodium benzoate E211), flavouring and antioxidant (Ca EDTA E385). Pea protein! Likewise, a house brand of peanut butter from one of the big South African supermarkets includes peanuts, sugar, vegetable fat (palm kernel), hydrogenated vegetable fat (canola, soya bean, cotton seed, TBHQ) and salt. Just about everything we know to be bad for us and our planet – added sugar, salt, palm oil products and hydrogenated fats – is in there.
[2] *Swallow This: Serving Up the Food Industry's Darkest Secrets* by Joanna Blythman (Fourth Estate, London, 2015)
[3] https://www.nih.gov/news-events/news-releases/nih-study-finds-heavily-processed-foods-cause-overeating-weight-gain (accessed 8 September 2020)

sodium nitrate, potassium nitrate and such being added. And those are still the relatively okay ones.

The food additive that has had by far the worst press but, somewhat ironically, is not at all harmful for us is monosodium glutamate, MSG.[4] It occurs naturally in food such as cheese and tomato. The Japanese started making it from seaweed broth in the early years of the 20th century to add that elusive fifth taste element, umami, to foods.

The bad press for MSG started in 1968 when a doctor wrote to a medical journal to say that he sometimes felt ill after eating Chinese food. The good doctor noted that since MSG was widely used in Chinese cuisine, mainly in soya sauce, it *might* be the cause. And thus began the "Chinese restaurant syndrome" and branding of MSG as a villain in our food chain.

Astonishingly, this mere suggestion has led just about all of us to view MSG with suspicion still five decades later. This is but one example Blythman cites (along with placebos, homoeopathy and anti-vaxxing) to demonstrate how deeply susceptible we are to mere suggestion and yet often resistant to solid data-driven information. There might be an evolutionary advantage to all of this, but it's hard to imagine what it might be.

Food manufacturers aim to produce the most they can at the least cost to themselves: Starting with a blank page, anything and everything is possible. It is, says Blythman, a never-ending chemistry lab experiment using both natural and manufactured food ingredients that are deconstructed right down to molecular level and then re-assembled in new and exciting ways.

A lot of food, or what goes into food, is perfectly legal, but that does not mean it is good for us – or is even food as we understand it. Food producers and their suppliers discuss "flavour technology systems" and "flavour delivery systems" rather than actual food.

---

[4] The generally respected US Food and Drug Administration "considers the addition of MSG to foods to be 'generally recognized as safe' (GRAS). Although many people identify themselves as sensitive to MSG, in studies with such individuals given MSG or a placebo, scientists have not been able to consistently trigger reactions." https://www.fda.gov/food/food-additives-petitions/questions-and-answers-monosodium-glutamate-msg (accessed 17 August 2020)

Take butter, for example. Did you know that what passes for butter in mass-produced baked goods and some dairy products contains as little as 0.02 per cent butter *extract*?

Among the products trafficked at food manufacturing fests are Butter Buds: "an enzyme-modified encapsulated butter flavour that has as much as 400 times the flavour intensity of butter".[5] Butter Buds cost only a fraction of real butter and do not need to be refrigerated. In many of the factories that produce these kinds of products, the workers have to wear protective clothing. This is heavy industry and nothing like what you would have seen in your grandma's kitchen.

Packaged meats, cheese, fruit juices and breads are seldom untainted by the food scientists' dabbling. Some contain only a small amount of the thing you think you are buying. But is it all bad? According to a fact sheet put out by the South African Allergy Foundation, "Without additives and preservatives a great amount of food on shop shelves would spoil before being bought."[6]

That much is true – but therein is also embedded deception by omission. The fact sheet continues: "The vast majority of additives and preservatives *appear to be* safe for *most* people. Laboratories throughout the world have tested them before they are used in foods. However, some individuals *may be* sensitive to various additives and preservatives."[7]

In South Africa it is primarily the Department of Health that is responsible for regulating what is and what is not allowed to be added to our foods, and how much, including pesticides and other contaminants.

Buried deep inside the pages of the relevant government "regulations and standards report" is that the measures given are a "daily recommended amount" that should not make you sick or cause you to die.[8] But what if you happen to have a sensitive constitution

---

[5] Joanna Blythman cites this strapline for Butter Buds on her blog http://www.joannablythmanwriting.com/journalism/inside-the-food-industry---the-surprising-truth-about-what-you-eat.html (accessed 8 September 2020)

[6] http://www.allergyfoundation.co.za/wp-content/uploads/2016/11/22-food-additives-004.pdf (accessed 17 August 2020)

[7] *Ibid* (my emphasis)

[8] Republic of South Africa, Food and Agricultural Import Regulations and Standards (Government Report, 21/12/2017). Also see https://agriexchange.apeda.gov.in/IR_Standards/Import_Regulation/FoodandAgriculturalImpor-

and you help yourself to a second serving of Farmer Joe's herbicidal broccoli? In that case, you get a double dose of his poison along with your daily portion of greens.

When it comes to processed meat, the less you know, the easier (but not necessarily better) it will be to swallow. The beginning of the problem is that meat is tricky to preserve. Biltong is one way, but even most of the billies we buy now has been treated with homogenised and mass-produced nitrates and nitrites. If ever you wonder why so much biltong tastes much the same – and, frankly, pretty bland – these days, that's why.

A visit to the website of Crown Foods,[9] one of the country's largest food ingredient suppliers, had my eyeballs popping: "With the ever-increasing price of fresh meat, we at Crown National offer products that enable you to achieve the best possible yields from your fresh meat. Our Batch Packs system combines food science, sensory technology, and analytical chemistry to ensure that the Boerewors, Bangers or Burgers that you produce will achieve the maximum profit margin at the retail counter." This is not information they mean to share with the general food-buying public but rather is aimed at the people who make our food.

Who would have thought the wors you just threw on the braai was the product of the latest analytical chemistry to give the flavour of (whatever it says on the label)? And what about food additives, functionality, sensory technology: Is this really still food we are discussing?

In modern food production, there are many substitutes for the real thing. For example, there are various manufactured proteins that are used to replace or mimic natural fats. Another classic example is egg. Seeing "egg" on a food label should not lead you to think an actual egg was cracked into a baking bowl and then mixed into your pie or custard or ice cream or mayonnaise.

Eggs used in processed foods are delivered either in powdered form or as a roll that can be conveniently cut into exact portions by machines. Or it could be "egg replacer" made from fractionated whey protein

---

tRegulationsandStandardsReportPretoriaSouthAfricaRepublicof4112019.pdf (accessed 8 September 2020)

[9] http://www.crownnational.co.za/ (accessed 19 August 2020)

(shelf life 18 months). Have you ever wondered how your supermarket or petrol-station pie got that glorious golden-brown crust? By adding synthesised carrot extract, just one of many "extract" ingredients that mimic once-real products, from rosemary to banana, is how.

It's the same stuff that gives yellow colouring to long-life custards, fruit yoghurts and numerous other products including sauces, salad dressings and mayonnaise. As for that freshly baked pie or croissant or bun: It might have been freshly heated, but the making was done somewhere else and came with an instruction, something like "heat for eight minutes at 120°C, then three minutes at 100°C". And that's before any synthetic topping is added.

The biggest con in the food industry is bulking, although it is mostly harmless health-wise. The magic dust of bulking is simple starch. It can stand in for up to 90 per cent of fats, 30 per cent of creams and yoghurts, and 25 per cent of tomato and other pastes or mayonnaise, and can totally replace eggs.

Your yummy, traditional double-cream Greek yoghurt might as easily get its unctuous texture from starch as from dairy products or yoghurt cultures. Just about whatever texture you are aiming for – creamy or crispy, crunchy or airy, gummy or elastic – you can get starch to do. "It is the muse of the modern food industry," writes Blythman.

Starch kept the food boffins busy for a while. But then came a versatile suite of food imitators that have been likened to nuclear warheads in the arsenals of ingeniously engineered food alternatives. These are the kind of enzymes we previously encountered in things such as dishwashing liquids, and they're used to make paper, tan leather – or fade designer jeans. I read in the *Daily Maverick* on the day I gave this chapter a final read that Israel had forbidden Heinz from calling its tomato sauce in that country ketchup, because it contained almost no actual tomatoes.[10]

These days, enzymes are found in just about every food you buy other than fresh produce – chocolates, sweets, chips, breads, cereals (including whole grain health ones), cooldrinks, beer, wine, cheeses, jams, animal feeds, pet foods… Are they bad for us? No one knows – yet. There are around 150 of them currently in use but, since they

---

[10] www.bbc.com/news/business-34052147

are not ingredients in any definable way, you won't find them on any food label. When people tell you they have a food allergy, it could be enzymes or colourants.

This new wave of food additives has bred another new monster, the spin doctors who are paid to make it all sound yummy. Among the terms devised to replace the old frightening chemical formulas include high-fibre, low-calorie, antioxidant, gluten-free, sugar-free, "high in" or "low in" something. Low fat is among the biggest food faux pas of all time.

In the United States in the late 1970s, research linked *some* fats to health issues, most specifically heart problems and obesity. It seemed to make sense: Eat fat, and you'll get fat. But it wasn't that simple, and it has cost us plenty.

Influential research produced in the 1970s by esteemed US institutions such as Harvard and New York universities, suggested[11] that if we reduce our fat intake, we could reduce our body fat. The actual science was much more complicated than that, but the fine details were deemed too complicated to feed to the masses.[12]

In the rush to catch this food fad bus, nutritionists and even some influential medical people told us fats were bad but carbohydrates (breads, cakes, anything with sugar which broke down to simple sugars when digested) were good for us. After that, anything low fat was punted as a wonder food, and total food fat consumption in the United States went down – yet it appears that the nation as a whole kept on getting fatter.

Correspondingly, the French, a nation known for a diet based on fats – butter, goose dripping and duck fat, fat-rich sauces and butter-rich *patisserie* – seemed to remain remarkably trim and healthy. What was going on?

---

[11] *Eat, Drink and Be Healthy: The Harvard Medical School Guide to Healthy Eating* by Walter C Willet and Patrick J Skerret (Free Press, New York, 2001). Also, *Food Politics: How the Food Industry Influences Nutrition and Health* by Marion Nestle (University of California Press, California, 2013).

[12] https://www.statnews.com/2016/09/12/sugar-industry-harvard-research/ (accessed 8 September 2020) https://www.npr.org/sections/thesalt/2016/09/14/493957290/not-just-sugar-food-industry-s-influence-on-health-research (accessed 8 September 2020)

Professor of epidemiology and nutrition at Harvard, Walter Willet serves a slice of sanity in an interview published in 2004: "This campaign to reduce fat in the diet has had some pretty disastrous consequences. We can very easily get fat from eating too many carbohydrates; we really have to keep an eye on calories no matter where they're coming from."[13]

Fats are essential for our health, facilitating the uptake of vital nutrients in our cells. What we need are the good fats, like those contained in olive oil and coconut milk, avocados and – yes – meat, and especially omega-3 fatty acids that come mainly from fish.

What nobody could have anticipated was the profound way in which the food industry would jump on the disinformation and replace (good but expensive) saturated animal fats with (mostly bad but much cheaper) unsaturated and polyunsaturated vegetable fats and sugars. A strange anomaly is how slow some nutritionists have been to admit the big fat blunder, even the South African Heart Foundation, which still endorses that slime of modern nutrition, margarine,[14] which the spin doctors have renamed "vegan fat spread".

A problem for food manufacturers is that unsaturated vegetable (liquid) fats are annoyingly unstable and hard to incorporate into foods. In order to make them more stable, suppliers did something called hydrogenation, the same process used to harden soap. That worked really well, and very soon we found hydrogenated polyunsaturated or "trans" fats in just about all baked and crumbed products.

All that sticky stuff on floors and surfaces where fast foods are prepared: It is not actual oil but chemical polymers that are by-products of cooking with trans fats. When it was found these kinds of fats were extremely bad for us (as much as for the palm plantations formerly known as rainforests) and were linked to heart disease, strokes, diabetes and probably cancer, the first response of the industry was what it always is: deny, deny, deny.

---

[13] https://www.pbs.org/wgbh/pages/frontline/shows/diet/interviews/willett.html (accessed 19 August 2020)
[14] www.heartfoundation.co.za/productscategories-category/margarine-reduced-fat-spreads/ (accessed 8 September 2020)

But you can conceal a deadly lie only for so long. Trans fats are just about the worst thing you could eat besides arsenic.[15] The big problem with eliminating fat from food was that the end product was bland; the fat is where most of the flavour compounds reside. Strip out the fat, and food manufacturers had to replace the natural flavours with something else. That something was sugar, which has subsequently become the world's number one nutritional terrorist. The quantities going into our foods these days are staggering.

Even so-called health drinks and flavoured waters are packed to the gunwales with sugar, often combined with caffeine. That's okay for young athletes but not so good for the spectators, especially ones on the couch. Even tonic water, which you'd think was bitter, has more sugar than a brightly coloured fizzy drink. We now know it is sugar that has made the world a very much fatter place – but we're addicted.

The *British Medical Journal* labels sugar as public health enemy number one.[16] And so, we have flirted with beet and agave syrup, stevia, xylitol and fructose (corn) syrup. But it transpires that some artificial sweeteners might be even worse for us than cane sugar. Variously, they have been linked to gout, liver disease, hypertension, diabetes, the dreaded Big C and even obesity – the very thing they are supposed to counteract.

Our smart brain can detect these deceptions. When it senses it's getting sweetness but without any of the anticipated calories, it lures us on a feeding binge to make up the shortfall. But why do we crave sugar so much? It is the fuel that powers us: If our blood sugar levels drop below critical levels, and our brain is starved of sugar for only a short period, it basically shuts down. Ask a person living with diabetes to explain.

We seem to be programmed to crave sugar, salt, fat and protein – things essential to us but hard to come by in nature. Too little, and we die. Too much, and we also die. They are the things our indigenous

---

[15] Used in small quantities, arsenic was long used to treat serious bacterial diseases, including syphilis. In fact, what with the spread of antibiotic-resistant bacteria, it's a treatment that might enjoy a comeback.
[16] https://blogs.bmj.com/bmj/2014/06/30/the-bmj-today-sugar-public-enemy-number-one/ (accessed 8 September 2020)

people, the Khoi and the San, coveted above everything else but water.

Now that these things can be had at just about every corner we turn, we don't seem capable of controlling our hunger, whether it's finishing the big bag of chips or guzzling one more ball of Swiss chocolate. Our brains are telling us to stockpile because there might be lean times just ahead. Except, these days, there is virtually no chance of that. *Bon appetit.*

Food for Thought II
The Broken Chain

AT A PUBLIC TALK NOT so long ago, television chef and good-eating champion Justin Bonello asked the audience how many among them had, has had, or knew someone who had cancer. Most hands in the large (albeit mostly elderly) audience went up. It did not used to be like this, did it, he posited. No, it did not, they almost all agreed.

Some might point out that we used to die mostly of old age, but now it's things like cancer and coronary thrombosis, because we have become so acutely aware of them, and good at diagnosis and data gathering. Maybe so, but there might be another reason, or at least a major contributing factor, and that is all the bad stuff that is being put into much of the food we buy. A lot of it should be making us feel sick – and it is.

Our pets as well. Decades ago, a vet would hardly ever think to look for cancer in a sick domestic animal. Today it seems it is among the first things the vets look for – that and high blood pressure as a result of eating mass-produced foods that are very bad for them. If you think about what goes into the very worst of human food, whatever is left over from that goes into Pookie's bowl.

Justin entertained the crowd with tales of growing up in Cape Town, a quintessential urban boy, riding bikes with his mates, stealing the occasional apple and getting into all kinds of nonsense, as any inner-city kid will. Back then, you played in the same street, or close by, where Mom bought fresh food from the greengrocer on the corner, and if he saw you getting up to mischief he'd call your mom, and when you got home, your dad would give you one.

It was a time before sunscreen was invented, and you came indoors when you were hungry (if there was blood, you washed it off before the folks saw it and gave you what for). Where I lived most kids had silkworms, and your best friend had a mulberry tree in their garden (get the juice on your school uniform, and your mom would give you another one). We watched bean sprouts grow in wet cottonwool and avocado pits – suspended with matchsticks in a glass of water that our moms put on the sunny kitchen windowsill – grow roots.

If you are around 50 or older and grew up in a large town or city, there was usually a chain of command, whether it was District Six or Gardens, Bryanston or Belville, that included your neighbours, the grandparents on your street, the postman and the grocer. That's how things worked back then, and they worked pretty well. Then we blinked, and it all changed.

I remember a place up the river from us called Tip Top Farm. We knew our fresh food came from there, as well as some exotic things they got from their Portuguese connections in Mozambique, such as cashew nuts, litchis and coconuts. Then one day, it was gone, and bulldozers were carving out a new suburb that came to be called Glenadrienne. With that, our local greengrocer disappeared, and so did a way of life. We no longer knew where anything came from.

Next to go was the green belt and the pig farm down the river. They got incorporated into Petervale, along with our childhood wilderness. That's how it's gone for maybe as much as half the world's population alive today.

Justin talked about the failed food system. "This is the broken chain of our modern lives," he told his rapt audience. We no longer know where our food comes from, or even what it is. The old neighbourhood system is broken, is gone, he said, and you could see the old minds in the hall recalling better times, when you went shopping for everything you needed in the main street wherever you lived.

In *Swallow This*, Joanna Blythman outlines how we are led to believe that what goes on in food factories is essentially the same as home cooking, only on a much bigger scale. Essentially, if you have seen the TV adverts, that pixies make our biscuits, and someone's grandma makes commercial ginger beer.[1]

Having been a food brand ambassador for a major food retailer and walking the talk for a number of years, Justin has now adopted a completely different position. He told the audience that the "free range" thing is usually a smokescreen. If you were to see the places where supposed free-range animal products are produced, you would not see anything that looked free, or anywhere that looked like a range. He argues that the definitions allow for just about anything: So long as an animal can walk, or at least shuffle, the law considers it to be "free range".

When you read the small print on the back label – using our number 2 reading glasses – you'll search far in supermarkets to find free-range chicken products that do not tell us they are "not regularly dosed" with antibiotics. A healthy, genuine free-range chicken should *never* need to be dosed.

Our family changed its eating habits ever since watching an episode of *Carte Blanche* around a decade ago where they investigated the dairy farming business in South Africa. It showed milk cows staggering with udders the size of beach balls. They found pus in some of the milk they tested. Places where cows were so ill, they had to be given regular doses of antibiotics. We are now a family of obsessive food label readers.

It has been argued, by medical professionals among others, that the abundance of growth hormones that comes to us courtesy of the modern food production line could be a contributing factor to girls in the developed world starting to menstruate much earlier than was previously the case (as young as seven).

After we watched that *Carte Blanche* episode, our family vowed to buy not only dairy products but as much food as we could that was certified free of all the bad stuff. Life seems a whole lot better now.

The advent of chemical fertilisers, pesticides, herbicides, large machinery and cold storage has revolutionised food-delivery from farm to city. Delivering huge amounts of fresh food to distribution warehouses, and from there out to the retailers, would be impossible

---

[1] *Swallow This: Serving Up the Food Industry's Darkest Secrets* by Joanna Blythman (Fourth Estate, London, 2015)

otherwise. Like water-borne sewerage, this food delivery system is one factor that has facilitated the growth of modern cities. But each has led to its own environmental issues: Most of the costs are carried downstream, by us as well as by the natural environment.

It's good for business without a doubt, but it's starting to kill us in ways we did not anticipate when modern industrialised farming really took root after World War II. And even in some cases where the harm is known and well documented, there is a culture of denial. This extends to the killing of people who have stood up to "big farma".[2]

It's hard to say exactly when the post post-industrial health food movement started in South Africa, but the name Jeanne Malherbe often comes up when the subject is discussed. Back in the 1960s she started farming according to the biodynamic methods espoused by anthroposophy. It was part of a larger movement based on the philosophies of Rudolph Steiner, who was an Austrian clairvoyant, philosopher, social reformer, architect, economist and esotericist, according to Wikipedia. The Waldorf school system is part of this philosophy.

Among the steep, pleated, fynbos-covered foothills of the Hawequas Mountains above Wellington in the Western Cape, Malherbe started doing things that had the locals raising eyebrows and giving one another the knowing "Boland buk" (a small, almost imperceptible head nod). Here was a woman, farming, on her own, and doing it a bit funnily, bio-dynamic-ally.

It's basically organic growing but with some esoteric elements added. One thing that – at first, to newcomers – appears to be part of the esoterica (like planting or harvesting at night, under a full moon, maybe even breaking out into a dance), is stuffing a cow's horn with manure and burying it in a new field. Except, it's not esoteric at all: it's good old-fashioned common-sense soil husbandry.

The bio matter inside the horn is set upon by all manner of microfauna and flora, those microscopic organisms that do the heavy lifting in creating healthy humus, the stuff that is the bedrock of healthy farming. It's a kind of super compost, similar to earthworm "tea" used

---

[2] In 2017, there were 207 recorded killings worldwide of food and farming reform activists. See https://bit.ly/2QmfGiC (accessed 23 August 2020)

by many small and home growers these days.

Malherbe allowed interested people to join her in what developed into a kind of meditation centre for lost and searching souls. Slowly, as the movement gained a foothold, so too did the biodynamic farming method. Today, numerous farmers, including a small corps of winemakers in the Cape, have converted to the biodynamic way and, if it works, don't knock it.

Who was first is impossible to determine, but in terms of alternative food markets in the city, the Bryanston (Waldorf) Organic & Natural Market in northern Johannesburg certainly was a game changer. Along with organically and biodynamically grown foods, the market has been offering various artisanal crafts based on a "not harming nature" principle for 40 years. They set a high bar for the community markets that have now been started around the country in places as far flung as Fish Hoek, Sedgefield and Howick.

Of course, Cape Town, being stuck way out in the Southern Ocean, has always felt a little otherwise. Back in the 1980s, now-luxury suburbs such as Hout Bay and Noordhoek had parallel barter economies. Today the Cape Peninsula has as many local markets as it does tattoo parlours.

Prince, if not king, of Cape Town's good food markets has to be the Oranjezicht City Farm. It was started in the first decade of the new millennium on a neglected bowling green in the embrace of Table Mountain, where one of the original homes of the City Bowl once stood, domicile of the Von Breda family. In 2012, some local citizens successfully lobbied the council for the park to be turned into a food garden based loosely on the allotment system.

Since then, Oranjezicht City Farm has blossomed into an all-encompassing community project, bringing together people from many walks of life and as many economic sectors and precincts. If there is a blueprint for urban farms, this has to be it. It is a truly magnificent piece of city renewal with a purpose and an inspired example of garden landscaping. Produce is sold at Granger Bay, adjacent to the V&A Waterfront, on Saturdays year-round.

But there is another initiative that far eclipses these markets in both size and vision, and it's pretty much all the vision and work of one man: the self-same "cooked" Justin Bonello.

I call him one fine spring morning, and we arrange to meet at his Neighbourhood Farm in – unexpectedly – rather dreary Fish Hoek. Even more unexpected is to find the market garden and store located in the grounds of an equally grim-looking state institution, False Bay Hospital.

When I get there, Justin is busy-busy but, between tasks, manages to inform me that it was his grandmother who helped him turn a childhood fad of growing food plants into cooking. He still has the crêpe pan she gave him. Memories of cooking with his *ouma* in the family kitchen and thoughts of today's broken food chain got him thinking: How could he get the children of his new 'hood on the south Peninsula to eat better?

Suddenly he gets up, grabs his car keys and, while he's headed for the door of his container office, he asks if I'd mind joining him on a visit to one of his food sites where input is required. It turns into a hyperactive two-hour, many-stop round trip.

"Modern city kids don't know where their food comes from," Justin rails. "Each new generation moves further away from its roots with the land. There is a disconnect and an increasing loss of traditional knowledge, and with that we buy into the prevailing urban snacks and fast-food system."

"They think the bread we buy is just bread," he says as we pull into Marine Primary, a school in Ocean View, a peri-urban township where the Coloured people of Simon's Town and other nearby white neighbourhoods were forcibly moved in the late 1960s. This is the site of one of Justin's five food-garden sites around the Cape Peninsula. "But it's not the bread we think it is. When you bake bread at home it has around four ingredients. These have around 20, and most of them you've never heard of." I've heard something like this before.

It takes some leaning in from me to get our conversation back onto my track, which is how he got the whole Neighbourhood Farm project going. It took years of pestering everyone he could think of. But the final key in the door was when he managed to get an audience with the premier of the Western Cape several years back.

He spoke to her about his childhood memories, the broken food chain and his vision of feeding the children well. He reminisced about growing beans and avo pits and good old stuff like that. He knew he'd struck the right chord, he recalls, when the premier opened the gallery on her mobile and showed him pictures of bean sprouts she was growing in a saucer of wet cottonwool.

From there he secured assistance from the provincial agriculture department. To start he needed vacant land and pointed out there was an abundance of it all around the town, like the hospital in Fish Hoek, as well as many of the schools themselves. In places such as Ocean View, for instance, some of the land designated as sports fields had reverted to weed lots.

"Children are products of their environment, and so are the teachers," my host says, with furrowed brows. We are standing in Marine Primary's impressive garden, part food and part ornamental. "People even come here to have wedding photos taken."

Of the nine schools where Justin got gardens started, only Marine Primary works autonomously. "It's all about getting the right guardian," he says, "but the hurdles are high. The schools should be taking over," he says, almost wistfully, but leaves off in mid-sentence.

Enter Elroy Kloppers, with his big Rasta beanie. Kloppers and I shake hands, and he tells me how he attended Marine Primary and was so taken by the garden idea, he returned as its guardian. "All" it needed was lots of water, mulch, compost – and clearly a lot of love.

The children here no longer kill everything they see on sight, Kloppers tells me with pride. "Education is our main activity here," Justin adds. "We have to teach the children to respect nature."

Back in the car, Justin tells me the one thing he has learnt is that water is the key to it all. "We have to be able to harvest rainwater from the gutters. If I see a school where the gutters are not maintained, I know it will be a dead loss and just move on."

On our way back, Justin points out the road works. "There is a broken food system," he says, "but there is also a general lack of understanding of systems, like what we do with stormwater, rainwater." Around 20 per cent of the urban landscape consists of roads where all rainwater gets flushed out to sea with whatever else it carries, as quickly as possible, the market gardener tells me. I tend to trust his figures, generally, because he seems like a man who does his homework.

Cape Town is notorious for wasting its rainwater during the long dry Mediterranean summers, having long ago turned most of its rivers into concrete sluices. "We are a semi desert, yet this is what we do with our rainwater," he says, pointing to the extensive stormwater sluices being laid in concrete on either side of the new roadway. "Surely there has to be a better way to manage our lives; we need to rethink how we live in urban environments."

Back at the high school, we walk over to a very substantial pile of waste. Mulch, I am informed: 1,300 tonnes of it. Justin tells me it comes from the nearby municipal waste site. A company used to charge the municipality to take it away, clean it up a bit, then sell it back to them as compost. "But I'm on to them," he smiles. A quick meeting with the garden manager, and we're off again.

Are all days like this, I ask. Pretty much, he says. "I had no idea of the size of the thing I was getting into: growing, retail, human resources, politics ..."

But for him, it's better than the alternative. In his previous life, when he was a brand ambassador, he met the Ferrari-driving directors and shareholders of food-producing companies and supermarket chains – people who are among our country's super-rich. At the other end of the supply chain are the farmers struggling to make a go of failing agricultural enterprises that have often been in families for several generations. And then there's the rest of us, the consumers, who are getting less and less nutrition for more and more of our rands.

Justin cites the example of Karoo lamb. A Karoo farmer typically gets between R28 and R38/kilogram of lamb, depending on the cut. For the same thing, you pay anything from R120 to R160/kilogram in the supermarket. There's at least R100/kilogram in there – three to four times what the farmer gets – probably going to the intermediaries and shareholders.

At the beginning of the new millennium, a farmer selling 200 lambs could buy a new bakkie. Today what you'd get from selling 200 lambs barely covers the deposit. In times of drought (like during the time this book was being researched) farmers are bent backwards over their feeding troughs and, it appears, beaten to financial death by the big retailers. In 2018, Justin notes, the top four food retailers in South Africa realised profits of around R17 billion. (I checked, it's close enough.)

"I don't want to drive a Ferrari," Justin muses while navigating his very well-worn Land Cruiser around the south Peninsula. "I want to get good food onto the tables of the rich, the middle class and the poor." All strength to you, Chauncey![3]

While there is a strong moral argument to be made that vegetarianism is the best diet for our own health, as well that of our planet, there are others who argue that "a modest amount of everything" is the best way to overall health.[4]

The real issue is where you buy your food. When you become a "locavore" and start buying local, it's almost always from small growers, producers and suppliers who take care in what they do and provide. By supporting local artisanal food producers, you also support small business and so strengthen local community links. Relinking the chain, I reflect.

This, in turn, kind of organically leads to more and more home cooking and less and less ordering in or eating out. Roaming your area as a locavore becomes its own self-rewarding adventure, especially on weekends, when it becomes a food and wine, or craft beer, outing. It's the new age way of being a hunter-gatherer.

There's no doubt that organically grown food is good for us, but it's also almost always more expensive than the typical "fruit & veg" offered by supermarkets, so for the majority of South Africans it is an unaffordable luxury. In chapter 10 we'll look at some alternatives.

---

[3] Played by Peter Sellers, Chauncey Gardiner is the lead character in the movie *Being There*: a simple gardener who deludes all the rich and smart people of Washington DC and is finally elevated to a kind of Zen serenity.

[4] See *The Omnivore's Dilemma: A Natural History of Four Meals* by Michael Pollan (Penguin, New York, 2006).

The Real Costs of Power
Execrable Coal, Fricking Fracking, Super Solar or Wonderful Wind

ONE OF THE MOST CHARMING books written by John Steinbeck tells the story of Frenchman Pippin Arnulf Héristal, who lives in a comfortable apartment in Paris, with his equally comfortable wife, Marie, and their gifted but dizzy daughter Clotilde. He loves nothing more than riding his scooter down to the boulangerie to buy croissants for morning coffee.

One day Pippin receives a call: The French people have decided they want a king again, and research has revealed he is a direct descendant of Charlemagne the Great. There are quite a few things Monsieur Héristal would like to put right, and so begins the short reign of Pippin IV.[1] But the new king finds life in the Palace of Versailles stifling (Marie not so much), and he also has virtually no power to do the things he hoped he might.

Now, imagine you woke up, and there was an urgent message for you: The Minister of Mineral Resources and Energy is out of action, and the country desperately needs you to take over that portfolio. They've been following your social media posts and reckon you are someone who really knows their wet coal from their hot gas. In short, you are the best person for the job. But be careful what you wish – or ask – for, because empires rise and fall upon their source of energy.

Stones powered the various Stone ages for tens of thousands of

---

[1] The short novel *The Short Reign of Pippin IV: A Fabrication* (Penguin Classic, New York, 1957) is a satire about French politics of the time.

years. Following that, timber fired up dynasties from the Levant, across Greece and Rome, China and India, until all the forests were gone. When the timber ran out, so too did the driving forces of those civilisations. Coal powered the steam age, which gave rise to the British Empire, and when that began to dwindle, it was oil that ushered in the rise of the United States as the world's powerhouse. It also elevated some previously marginal oil-rich nations to world players.

What would you do – given the fact that here in sunny South Africa we have no natural oil, no thermal energy sources, a little gas and precious little water to throw over falls in order to power turbines like they do in other, more watery parts of the world? Dear Minister of Emergency Measures, allow us to lay out the various options for you, as a basic, if brief, overview of the situation. Warning: It's not all – in fact, hardly any of it is – good.

COAL

First is the hard black stuff, our go-to energy source for the past couple of centuries. The only good news here is that its time and practical usefulness has all but run its air-polluting course. It's a good news-bad news resource: South Africa has lots of it, and it's awfully cheap: From my research and all that I have read, it's also some of the worst-quality coal in the world, with very high sulphur content. Most of it lies in the province of Mpumalanga, which calls itself the "cultural heartland" and which has caused that part of the Highveld to have among the worst air pollution and acid rain on Earth.

On opening Africa Oil Week in 2019, Energy Minister Gwede Mantashe (in a cherry red outfit) told the delegates: "South Africa has 16 coal-fired power plants and vast deposits of coal. We recognise that coal must disappear, which it will over time, but it won't be soon."[2] That does seem to be the honest truth of it.

A point that should not be overlooked is that Minister Mantashe is a union man until the day he dies, more specifically a NUM (National Union of Mineworkers) champion to his marrow. So, we should not be surprised should he defend those working the coalmines.

---

[2] https://www.dailymaverick.co.za/article/2019-11-05-gwede-mantashe-sticks-to-his-guns-on-coal/ (accessed 8 September 2020)

In a country endowed with some of the world's richest mineral deposits, it is a profound shame that we should have to endure regular power cuts because our national power utility, Eskom, cannot deliver.

The reasons they give for their non-delivery are woeful: A worker dropped a spanner into the works, a conveyor belt broke, our coal got wet, a dog ate our restructuring plan. In fact, the reasons are far more numerous and complex and include: a change from a meritocracy (your job based on being the most competent) to a mediocracy (anybody is capable of doing any job); a change to a kleptocracy; loss of institutional intelligence and memory; institutional incompetence; lack, complete or partial, of an infrastructural maintenance programme; state capture and the migration of capital; and union resistance to change.

One business analyst likened Eskom to a game of Jenga: The structure is teetering while our government pushes and shoves at it like the Addams family. All it needed was for one player to bump the table, and the entire organisation would come crashing down – followed by our entire economy.

But before you have us all leap off the coal wagon, consider the following. Currently around 90 per cent of South Africa's power and 40 per cent of petroleum fuel supply comes from coal. As Minister Mantashe put it: "People want us to switch off all the coal-fired generators." If that were to happen, he warns: "We'll breathe the fresh air in darkness."[3]

At the time of writing, 2020, as much as 40 per cent of the country's foreign revenue came from mining, and around 30 per cent of all mining revenues came out of coal mines. Those mines employ about 82,000 people (with Eskom accounting for another 50,000 jobs). In a country with an official unemployment rate of around 26 per cent, the loss of those jobs would see many more people on the streets. Jobs for coal; coal for jobs. Sit tight, and put on your gas mask – it's environmental carnage out there.

NUCLEAR

SPEAKING OF WAR, SOUTH AFRICA IS the only country to have ever voluntarily given up its stockpile of nuclear weaponry. We also have

---
[3] *Ibid*

Africa's only nuclear power-generating reactor, the famed, or ill-famed, Koeberg plant near Melkbosstrand, about 35 kilometres north of downtown Cape Town.

When it was planned in the early 1970s, Melkbos was considered a far outpost of the city, but now it's another outer suburb – not the best site for such a bellicose facility. Koeberg was envisaged to replace all fossil fuel power units in the province, with further plants planned for Ystervarkfontein near the mouth of the Gourits River and at Oyster Bay near Cape St Francis.

The environmental lobby put the brakes on the building of further nuclear stations. Meanwhile, the population of urban areas doubled, tripled and – in some cases – pretty much exploded. Between 1990 and now, Cape Town went from around 2 million to more than 4 million residents. During that time, the once extremely exclusive Plettenberg Bay went from a population of around 4,000 to around 60,000, it seems most of the new arrivals were unskilled people from the rural Eastern Cape looking for better prospects.

There is a view that this is good for the country at large, putting poor people closer to where they can expect to improve their lives, but we're discussing the power of energy here. Without a doubt, nuclear would have been the quickest fix here.

Environmental opposition and rising costs have rendered nuclear the most impractical of all our choices now, but still some persist. Word on the street is that the people in government still backing the nuclear dream are really after the kickbacks from ruble-toting Russians.

Green-movement opposition to nuclear power is not without good cause. In September 2010, some 91 workers at Koeberg found themselves setting off those sci-fi bleeping machines: They had been contaminated by the radioactive material cobalt-58. Since 2005 the nuclear facility has experienced numerous shutdowns and breakdowns that have caused economic losses estimated at around R2 billion,[4] and a lot of nervous people in the northern suburbs.

Along with coal, our leaders seem intent on not letting go of its nuclear bone. Two decades ago, an engineer would have said nuclear

---

[4] https://en.wikipedia.org/wiki/Koeberg_Nuclear_Power_Station (accessed 8 September 2020)

was our best option, based on output and best-/worst-case scenario risk assessment. You use up all the available isotope electrons then ship off the radioactive waste to Namaqualand, encased in steel and concrete, throw it in a hole and put up a sign: "Do not uncover, ever. Even you, Superman."[5]

Which brings us to fusion nuclear energy, the kind the sun gives us for free. Instead of splitting atoms the way fission power stations currently do, and leaving lots of highly toxic waste to deal with, fusion reactions harness the positive charges of atoms, leaving no radioactive waste at all. The problem is that it requires extremely high temperatures, as high as 15 million degrees Celsius, which is hotter than the surface of the sun. You have to make that temperature with some other power source, so back to square one and first principles. But maybe not…

Some clever scientists in Sydney have patented a process that uses lasers in place of raw heat to strip off the electrons of very common elements such as hydrogen and boron (at atomic level, speed equals temperature). This creates "naked" atoms (ones with no electrons), and the positive charges are used to set off the flow of electricity. Got it?

But sceptics in the know reckon fission is still where it has been for the past 50 years – always and ever 20 years off. We'll have to wait and see. Then again, if the Sydney patent works, we might even welcome nuclear reactors into our own back yards soon. The best and worst of it is that you cannot predict the future, which makes it so hard to plan for.

GAS

NOTHING NEW HERE, WE'VE BEEN playing with gas for a while already, ever since a parastatal named Soekor started drilling for oil in the Karoo back in the mid-1960s. Small traces of oil were found (as exploration geologists had predicted), but not enough to make it commercially viable. They also found gas, but in those days they didn't know how to extract it.

Later in that decade, Soekor went seeking for fossil fuels offshore. Around 90 kilometres off the southern Cape coast, they found quite

---

[5] South Africa disposes of its nuclear waste in a hole at a place named Vaalputs, way up north near Springbok.

a lot of natural gas, and so was born Mossgas, that industrial gateway to the region formerly celebrated as a garden route.

In 2002, as part of his great Africanisation programme, then President Thabo Mbeki's energy men repurposed it as PetroSA. The gas-to-liquid fuel capacity of the Mossel Bay plant has been reported as being around 45,000 barrels a day,[6] which is quite a lot and satisfies about 20 per cent of the country's fleet requirements. More recently, offshore gas deposits have been located in an exploration area named Brulpadda, about 175 kilometres off Cape Agulhas.

While all this was barrelling along, the world outside developed a new technology to extract those small pockets of gas trapped in Carboniferous rocks, what we call fracking – spit-spit, wash out mouth, all together shout "NIMBY!"

Around 275 million years ago, the Karoo was a vast and shallow watery basin covered with and surrounded by dense growth of primordial plant matter. When it died and was covered by sand and silt, the organic mass was slowly compressed, and the carbon content turned into gas, oil or coal, depending. The gas lies trapped in trillions of cracks in the southern wedge of the Ecca suite of rocks, while coal amassed around its northern arc.

The truth of the matter is fracking, or hydraulic fracturing, is just about the easiest and cheapest way of getting fossil fuel out the ground and into your car. In the United States, it has reinvigorated a flagging national economy and turned that gas-hungry nation from a net fuel importer to an exporter. What's not to love about it? To extract the gas, you need to sink wells and pump in a cocktail of "fracking fluids" under high pressure in order to open the cracks and release the gas.

Those fluids consist mainly of water, with sand or aluminium oxide particles in suspension to keep the fine cracks open. Added to that are thickening agents or proppants (stuff like hydrochloric acid, friction reducers, or detergents, biocides to prevent the growth of algae in the opened cracks, and emulsifiers such as 2-butoxyethanol) as well as radioactive tracer isotopes to follow what is going on deep down. What could go wrong?

---

[6] My man on the inside believes these figures are highly suspicious and exaggerated.

The typical fracking well will use an average of around 17 million litres of water during its operating life. That is water, in a dry region like the Karoo, that would not be available for human consumption or agriculture. Once used, it is seriously polluted stuff that needs to be stored somewhere. South Africa is estimated to have the eighth-largest shale gas deposits in the world (around 40 trillion cubic feet if it were bottled),[7] which would require a veritable sea-full of water to extract. But that balloon-busting amount of gas is hard for the fuel companies to ignore.

In South Africa, fracking could be a real energy and economic game-changer. In an ideal system, there would be stringent operation rules, one of which would be "you break it, you fix it, you pay for it". But South Africa as we know it is very far from being a perfect system. Even in the United States, the most litigious country on Earth, fracking accidents are legion, and recourse is very hard won.

When well seals are breached (and occasionally they will be), all those proppants and isotopes leach into the groundwater, and suddenly Farmer Koos's well water bursts into flames, Tannie Elsie's tap water spouts methane gas and other insalubrious goo, and the kids get sick. So, you can understand why the good people of the Karoo have pretty much all turned into NIMBYs – not in my back yarders.

The bottom line, however, is that you cannot refrain from fracking your gas and have it: Karoo farmers cannot expect to race around in bakkies rescuing sheep or shooting scrub hares and then demand that their beloved farmlands remain untapped, any more than SUV-driving parents can expect the air over their towns to remain pure blue.

The Karoo – around 60 per cent of our land area – accounts for a mere 20 per cent or so of our human population: Should its inhabitants be allowed to hold the rest of the country to ransom over what is, mostly, in reality, an aesthetic issue other than about farms going dry or gooey? Yes, probably, because the gains of fracking in the Karoo would only be short-term.

The Karoo – the arid plains, flat-topped koppies, farms and dorps – is our national heartland. It is also the spiritual heartland of San and

---

[7] https://en.wikipedia.org/wiki/Hydraulic_fracturing_in_South_Africa (accessed 8 September 2020)

Khoi people who felt the reverberations of the giant litho-gong surface through the soles of their feet, electrifying their veins and surging into their souls. And, as if we needed reminding, the Karoo Basin, and its multitudes of ancient fossils encapsulating 200 million years of evolutionary history, is one of the natural wonders of the world.

Fracking for gas would only ever be a short-fix solution to our energy needs, big and pressing as they are. Would we want to sacrifice the spirit of the Karoo, and its people, for a 20- or maybe 30-year reprieve for the age of fossil fuels?

When it comes to big-scale decision making, one has to consider best-/worst-case scenarios. In the case of nuclear, the best is very good, but the worst-case scenario is positively apocalyptic. A breach in a fracking well is not the end of the world, but it would be for a small part of it. Meanwhile, it turns out, another plot has been unfolding in the gas story, and it's set in Mozambique.

Our north-easterly neighbour has found huge offshore gas deposits in its northern Cabo Delgado Province. It has been reported in the *Daily Maverick* that over the past several years it has been exploited by mainly French (Total), Russian (Gazprom, Novatek) and American (Enron, and later our own Sasol) fuel companies.[8] While we were all obsessing over bakkie, booze and whether to wear masks or not, it seems our actual energy minister Gwede Mantashe has been laying a gas pipeline (not literally, of course) to the northeast. We have been a long-time importer of hydro-electrical power from Mozambique's Cahora Bassa Dam. Also, as far back as 2012, we helped build gas-fired power stations in Mozambique to send us vital power.

It might be no coincidence that Cabo Delgado (including some paradisiacal holiday islands of the Quirimbas Archipelago, such as moneyed playground Vamizi) has recently been invaded by a jihadist army from gas-rich southern Tanzania. At the time of writing, they had managed to route both the Mozambican military as well as a "proxy" private Russian mercenary force, Wagner, from the area.[9]

---

[8] https://www.dailymaverick.co.za/opinionista/2020-09-18-gwede-mantashes-gas-plan-the-northern-jihadists-and-russia-the-emerging-power-play-in-mozambique/ (accessed 21 September)

[9] https://www.dailymaverick.co.za/opinionista/2020-09-18-gwede-

This gas forms an integral part of our country's future power plan, and we have already invested heavily in infrastructure up there, according to African Climate Change Foundation executive director Saliem Fakir.[10] So, if we see South African ground troops and even our Navy[11] headed north soon, we'll know it is to protect our investment in cash and hopes in those gas fields. Maybe even the Russian Navy, given the increasing importance of gas to its own plans as well as its close ties to Pretoria.

It looks like the bakkie drivers of the Karoo won't have to cry NIMBY! after all, so long as the drilling for gas proceeds in our neighbour's garden up north.

RENEWABLES
AND SO, TO THE REALM OF the elements, namely wind and fire (sun), but also a little bit of water. While real affordable solar energy is a new thing and has become cost effective only in the past decade, people have been using wind to power ships, windmills and water pumps for thousands of years. Using straightforward technology, those wind turbines can deliver electrical power to your home as well as to the national power grid. Given that our coastal areas are virtual non-stop wind tunnels, you'd think we would have tapped into it a long time ago.

The costs of solar and wind power vary with time and also to some extent depending on which source you consult. I looked at various energy websites early in 2020 and on aggregate they put wind slightly lower than solar energy, which is probably why there are so many recent wind farm installations in the Western and Eastern Cape. While wind blows day or night, when it does, while the sun shines on us for only half a day at best. Either way, they are not reliable enough to power a nation. The issue is storing and converting nature's energy into electricity we can switch on and off at will.

To store energy when it is not blowing or shining on you, you need

---

mantashes-gas-plan-the-northern-jihadists-and-russia-the-emerging-power-play-in-mozambique/ (accessed 21 September)
[10] *Ibid*
[11] The jihadist force is largely a pirate one that uses boats to invade coastal areas.

big batteries. Just imagine how big they would have to be to store enough for the nation's needs. There are novel kinds of "batteries" being tested here and elsewhere, like the molten-salt solar tower that can be seen between Upington and Kakamas. Ours is a Spanish-pioneered technology that is part of South Africa's experimentation with various energy options.

Not everyone likes those impressively large windmills that are sprouting up around our coastal regions, and not only for their visual pollution of our lovely countryside. Bird and bat watchers fear the small creatures will fly into those giant armatures (the correct term for the blades, which are sometimes up to 48 metres long) and be carpaccioed. Or, that when they fly in the lee of the blades, their lungs will burst in the relative low-pressure zones. But hey, what's the price of a few blue cranes or sheath-tailed, slit-faced or leaf-nosed bats in exchange for perpetual power? (Fun fact: Bats are by far the most bountiful of all mammal groups.)

South Africa's electrical generation capacity at any one time is a little more than 51,000 megawatts (MW).[12] A typical modern wind turbine produces less than 1 MW per day (given that the wind doesn't blow all the time, or when you need it to). If we want to use wind to equal our current capacity, we would need more than 50,000 turbines to spin our various discs, heaters and coolers. That translates to one of those impressive 60-metre-tall wind turbines every 17 or 18 metres between Cape Town and Port Elizabeth, or one every kilometre, but 50 abreast.

Even if we did go down that road, there would still be a major problem, and it's called the base load – the minimum demand over a period of several days. South Africa's base load requirement is 2 gigawatts (GW). This means a large amount of power needs to be on tap at all times. With renewables, you get back to the extremely expensive business of power storage. And round and round it goes.

Currently, at the time of writing, around 90 per cent of our base load was being delivered by 16 fossil fuel power stations, mainly coal but also diesel in the wake of the wide-scale breakdown in coal supply, which is rather ironic given its abundance and low cost. The

---

[12] https://www.usaid.gov/powerafrica/south-africa (accessed 26 August 2020)

remaining 10 or so per cent is split between wind, solar, hydroelectric (mainly Drakensberg-Vaal, Gariep, Palmiet and Steenbras pump-storage facilities) and our one Archaean nuclear power installation at Koeberg.

Pump-storage or hydro-electric utilities are great because you use the grid power to pump water from a lower dam up to a higher one, usually late at night when demand is low. Then you run it back down through the turbines and into the lower dam when power is needed during the day. With coal and nuclear facilities, it's far too costly and complex to shut them on and off as power is needed, so they pump out power and pollution around the clock.

And we have to pay for all that base load, whether or not we use it. Needing power 24/7 are large industries and mining operations such as the coal-mining giant Exxaro, and the likes of Sasol.

These big guys use on average around 60 per cent of the country's total power output, and so they get special concessions from Eskom. Mining houses have been given up to 90 per cent discounts for 30 years. How does it feel to know you have been paying all those huge bonuses to their executives and holding up their share prices all these years?

In (marginally) good breaking news, the government's latest National Development Plan trumpets that, by 2030, our energy generation would be 60 per cent fossil fuels (coal, gas and diesel) and 40 per cent renewables (20.5 per cent wind, 8 per cent hydro, 7 per cent solar and the remaining 4.5 per cent nuclear).[13]

In an interview with Chris Yelland published as a series in *Business Maverick* in April 2020, André de Ruyter, who was appointed CEO of Eskom in 2019, confirms the power utility is looking at turning aging power stations into renewable energy hubs. But he cautions: "This is going to be a long passage as we wind down our reliance on coal and, unfortunately, I think we'll be a major coal consumer and hence a major emitter of greenhouse gases and associated pollutants for some time."[14]

---

[13] https://www.gov.za/sites/default/files/gcis_document/201409/ndp-2030-our-future-make-it-workr.pdf (accessed 26 August 2020)

[14] https://news.knowledia.com/ZA/en/articles/eskom-plans-to-convert-old-coal-mines-and-plants-into-gas-solar-and-wind-b836946c1e7c802cab84d409

De Ruyter says decommissioned coal-fired power stations could be converted to using natural gas, thereby saving between 30 and 50 per cent of the capital needed to build new facilities. Open-cast coal mines – a major environmental issue right now – could be rehabilitated into solar and wind farms. It would have sounded sweeter had he said "would" rather than "could".

All was well on track towards 2030, with wind and solar installations popping up all around the country. In 2018, government earmarked R56 billion for the development of renewables.

Then, in July 2019, suddenly and without warning, Karen Breytenbach, the head of the Independent Power Producers (IPP) association was fired by Energy Minister Mantashe. Since 2011, she had helped raise investments of around R200 billion for various projects and built the IPP into a world leader.[15] Thomas Garner, chairman of the IPP, told the press Breytenbach had been excellent in her job and was "absolutely incorruptible".[16]

Just look again at those figures, and you start to get a hint as to why she might have been shoved aside. It appears that many people fear her axing was to make way for more corruptible managers of the kind we have come to know so well.

Imagine if actual engineers – rather than politicians and their friends who have been steering our country's steam-age energy train for the first two decades of the 21st century – ran our power-generating utility. You can only imagine what deals were cut behind closed doors in order to hand over the steering wheel to De Ruyter, an actual former oil executive at Sasol and not ostensibly a party man.

If we look into the crystal ball, our most sensible scenario would be for each home or, at most, block of homes to be generating at least part of its own electrical requirements 20 or 30 years hence. By then, every new home should be built using solar voltaic roof tiles, and batteries will have become ever cheaper and more powerful (thank you Elon). But to me

---

27832edcaac39c49 (accessed 8 September 2020)

[15] The R200 billion Breytenbach raised was on top of the R56 billion earmarked in 2018 by national government for the development of renewables.

[16] https://www.news24.com/fin24/economy/axing-of-renewable-energy-office-head-a-major-setback-say-energy-experts-20190723 (accessed 26 August 2020)

it seems that people in positions of power don't like giving that kind of power to us, as it weakens their own grip on the levers of public control.

Dear Stand-in Energy Minister, you will find very soon that it is no easy job, what with all those backs and all those knives. We predict it will not be long before you too jump on your scooter and head for the nearest boulangerie.

FFS – The Dirty Little Company That Could
And the Green Bogeyman Who Couldn't

Do you remember Mr Wolf, in the movie *Pulp Fiction*, the character played so mesmerisingly by Harvey Keitel? Wolf was a man of few words but plenty of action. He was the guy the baddies sent in to clean up their messes and, if you did not work with him, the poo you were already in was going to get a lot deeper. This story is about the Mr Wolfs of South Africa's oil industry, the guys who are called in whenever there's bad stuff to clean up.

In the 1970s, if you went strolling along just about any beach, when you got home you'd most likely walk thick, black, tarry stuff all over your mom's carpets. You could scrub and scrub with soap, but the most effective way to clean it off your feet or the floor was with an industrial solvent or petrol.

That "stuff" was mostly waste HFO, or heavy fuel oil, used either to power ship engines, or crude oil from tanker ballast water dumped at sea. The thick oil would break up into globules and wash up on our beaches. This was how it was done back in the day. Times changed, and the International Convention for the Prevention of Pollution from Ships was adopted, and member countries are required to ensure their shipping companies safely dispose of their waste on land and not at sea.

Not all countries signed up for this, nor did all shipping companies, as there is a cost attached to responsibly disposing of shipping waste.

The business was revolutionised somewhat when waste oil became a commodity and shipping companies could make money from selling their waste, rather than dumping it at sea.

HFO, petrol, diesel, solvents like benzene and many other products are all fossil fuels, derived from four sources: crude oil, natural gas deposits, coal and, in more recent times, tar sands. What they all have in common is that they are composed primarily of hydrocarbons, with some other materials fuel engineers call "contaminants", things like trace metals, water and the occasional fossilised fish.

Carbon is the basis of all life on Earth and, if our laws of chemistry and physics work throughout the universe the way we think they should, then all life everywhere is really just stardust arranged around carbon skeletons.

The thickness or thinness, the viscosity, of a fossil fuel depends on the length of its carbon chain, the framework of its fundamental molecule, be it bitumen, tar, diesel or paraffin. You can use various mechanical and chemical processes to make the molecular chain longer or shorter, heavier or lighter, as you require.

All fossil fuels on our planet originated as living matter, specifically the plants that colonised land way back, between 360 and 300 million years ago, give or take a few million either way. All that carbon formed the backbone of the moss, fern-, cycad- and palm-like plants that formed great forests across most of the available land.

When these forests died and were covered with sediments, they were increasingly compressed to form carbon-rich rock strata. During the period we call the Carboniferous, what with so much carbon being encased underground, the oxygen content of the Earth's atmosphere was the highest it has ever been, around 35 per cent.

This high oxygen content allowed some land animals of the time, mainly insects, to attain great size: dragonflies with 75-centimetre wingspans, millipedes of 2.6 metres long, 70-centimetre-long scorpions and giant cockroaches. An entomophobiac's nightmarish Carboniferous Park. It is also what prompted the dinosaurs to get so big.

But now that we have been exhuming and burning fossil fuels at an accelerating rate for the past 200 years, carbon in our atmosphere has risen, and oxygen has dropped to around 21 per cent. That, in its chemical essence, is what our troubles with climate change are about.

On 6 October 1973, one the eve of Yom Kippur, the most sacred of Jewish holidays, a coalition of Arab states led by Egypt and Syria attacked Israel.[1] After a bloody three-week conflict, Israeli forces managed to drive the Arab armies back on two fronts and occupied Arab territory in both Gaza and Golan, deeply humiliating the Arab invaders.

In 1974, in retaliation and reprisal for their defeat, Arab oil producing nations placed an oil embargo on those nations they considered to have supported Israel, particularly the United States of America, but also South Africa. (Remember, we got a lot of our arms from Israel back then.) It caused not only a worldwide oil crisis but also a mid-sized economic meltdown.

Business people will tell you that, when you see blood in the streets, you should buy stock. In Durban, an entrepreneurial engineer named Tony Hurter sniffed trouble in the air and jumped in. Not into a pool of blood, but rather one of oil. Waste oil, to be correct: sludgy, heavy, thick, contaminated waste oil. For around 40 years prior to that, a consortium of the oil companies BP and Shell had been pumping waste from their oil tanks at the refinery in the Isipingo area south of the Durban Bluff onto open land adjacent to the old airport.

If left alone, the substantial oil swamp would not have become an environmental catastrophe (it did not infiltrate the groundwater or seep towards the beach), "although it did kill some ducks", according to someone who was there. Ironically, given a healthy dose of nitrogen as part of the oil parcel, the grass grew better there than anywhere else around the old refinery.

But, if anyone tried to meddle with it, the toxic lake would almost certainly become both legally and environmentally delinquent and would prove to be a very big problem. Eventually someone would have to, but that would be someone else's problem, at some future time. That's how the executives in big companies, like big oil companies, often think.

Anyway, the someone of this story was Hurter. He looked at the

---

[1] en.wikipedia.org/wiki/Yom_Kippur_War

big oily swamp lying south of the harbour and sensed a business opportunity. He and his wife, Diane, floated a new business with the intention of trucking the big pile of sludge back to their HQ in the Jacobs industrial park not far away, spin it and scrub it clean, and then sell it back into the fossil fuel market.

The local whaling industry located on The Bluff had only recently closed, so the Hurters hired the general manager of the Durban Whaling Company – the primary product of which was, of course, whale oil – to help them run the new venture.

They named it FFS, Fuel Firing Systems, although some people have voiced another possibility for the name. From their core business of unloading waste oil from ships in Durban harbour, they had learnt how to use centrifuges to clean out the sludge and seawater in the ship's waste fuel. Later, they would also acquire old car oil, and many other spent fossil fuels down the line, and do the same.

In time, using the Swedish Alfa Laval technology, they found they could take just about any carbon waste product and repurpose it for just about any client with a fuel need, such as Consol Glass to melt quartz rock and make glass, or confectioners to bake breads and buns.

Each barrel of waste fossil fuel repurposed means one fewer barrel of crude oil needing to be dug out of the ground, shipped to a refinery somewhere, and then processed with all the attendant pollution and risk.

In the process, FFS was not only the pioneer in cleaning up waste oils on a large scale, but for a while was the largest user of Alfa Laval centrifuges in the world (up until the business of recovering oil from vast tar sand deposits in Canada, and later in the United States, boomed; you have to filter out all that sand and bear poop).

Sasol was another business that caught the Hurters' attention. It had accumulated a large… actually a very large pool of what was euphemistically called "unutilised hydro-carbons", which the company was leaving for someone else, at some future time, to deal with (read: us, the ever-suffering masses).

But here again, FFS was not averse to putting on its metaphorical HAZMAT suit and gumboots, and getting stuck in. What it jumped into at Sasol was around 6 million tonnes of coal fines (dust) being stored in sludge dams, as well as nearly 2 million tonnes of other bitumastic waste products, also stored in holding dams.

FFS brokered an agreement with Sasol to suck the muck out of the dams, send the feculent matter through its centrifuges to remove non-combustible stuff, mostly metals and ash, and then sell the cleaned up heavy fuel oil to clients in the shipping and other big industries.

Then the company would buy the waste back from the ships, again clean it up and break the heavy, long carbon-chain, bitumastic oils into lighter, less viscous ones – such as waxy oils, petrol, paraffin and benzine – then sell these to glass or brick or bread makers, and so on. It was (and still is) a dirty business, but someone's got to do it, or else we'd all be in an even deeper fossil fuel cesspit than we already are.

Aging refineries such as the old Caltex/Calref in Cape Town can typically recover only around 60 per cent of the contents of a crude oil barrel into useable fuels. More modern refineries can get a full 100 per cent of usable, mainly light, product for your barrel buck, mostly in the form of petrol and diesel.

The obvious solution would seem to be to close down all the old, inefficient refineries, and in truth there are moves around the world to do just this – but that could have catastrophic consequences for local economies and even some entire island nations, like New Zealand.

As with all things big and business, the move to close old refineries is being driven by oil traders who can get two or more times their investment back from selling lighter fuels such as diesel and paraffin, with nothing in the way of heavy fuel oils for ships left at the bottom of the barrel.

The shipping industry has been pretty much keeping its head, maritimely speaking, in the barrel about this looming economic iceberg, because it will require it to change from heavy oil to diesel engines. That will be no bigger a mechanical deal than changing a few injectors and pipes, but the fuel will come at twice the cost – or more. This fuel swap will happen, and those in the trade believe it will be sudden when it does.

And then, equally suddenly, ships will not be able to afford to sail to far-flung islands like New Zealand. And when they don't, it and other little islands will not be able to export or import anything other than that which comes by plane or sea currents, like coconuts.

But hey, that's their problem, not ours, and we've got enough of our own, okay. Like what closing down the old Caltex (Astron Energy) refinery in Cape Town would do to the local economy. The refinery on the N7 is one of, if not the largest, employer, directly and indirectly, in the southwestern Cape region.

The Western Cape is a major exporter of agricultural produce; the entire province relies on its agricultural sector for the majority of its economic activities (deciduous fruit, grapes and wine, wheat). It also ships out iron ore from Saldanha Bay, mainly to Japan. Double or perhaps even triple the costs of exporting by ship, and the crown jewel province goes belly up.

The alternative is, mainly, to source new markets within Africa, which would not be an altogether bad thing. Also, it would reduce overall pollution, what with no more bananas coming in from Guatemala, no more out-of-season avocados from Israel, no more American chickens, no more fish from Scandinavia, no more whisky from Scotland, no more beef or nuts from Brazil, and no more Oreos from Bahrain. (Yup, that's where they come from.)

Knocking on, it would mean less importation of crude oil that, at around US$17 billion a year (prior to the COVID-19-induced slump), in early 2020 constituted around 20 per cent of South Africa's imported commodities. Overall, we would have to do with less. We would also be forced to make, grow and buy locally. There would be a net decrease in CO_2 emissions and a net increase in local commercial exchange.

What will more likely happen, though, with lots of argy-bargying, lobbying and horse-trading behind the scenes, is that HFOs will not be scrapped altogether. Old refineries will be given a remission to clean up their act, modernise, reduce emissions – and get with the global programme. FFS has already shown it can be done and also how to do it.

However, they, the refineries, will do it only when the HFO gun is held to their head. It's just the way the people who run these things operate. After all, why clean up your act just to make the world a cleaner place for our children? Those old refineries were not designed to be clean, or to produce clean products, but they can be.

Then again, total closure would also drive us to implement immediate alternative energy systems, and that would not be a bad thing, in the long term.

But not all the FFS stories were grim ones. A funny one, in retrospect, concerns the pipe they noticed coming out of a brand-new Sasol facility.

"What's that for?" an FFS person on site asked a Sasol person.

He had no idea, other than it was for some waste by-product. At the very last minute before turnkey time, Sasol begged FFS to loan them a container tank, any old tank, as a matter of utmost urgency, to put under the mystery pipe.

FFS had only one old, rusting tank to spare, so they cleaned it up as best they could and had it delivered and installed.

"What! You can't give us that! You'll have to paint it."

So, they (FFS) did. Much later, when hardly anyone could remember anything about the incident, FFS got a message from Sasol complaining about the "unsightly" tank they had supplied and more telling than asking them to remove it pronto, or else.

"Tony Hurter was a really great salesman," recalls someone whom he recruited to expand the company's technical capabilities. "He would find stuff that other oil businesses did not want, basically their waste, offer them money for it, clean it up, and then sell it to new customers at a great price."

Like when he discovered Sasol had heaps of waxy oils they could find no use for nor get rid of. FFS bought the lot, around 3.5 million tonnes in all. "It looked like grey shoe polish. But we found that if we warmed it up first, it could be used in various heating processes that had previously used other thinner products," according to the former FFS employee.

Sasol also had two very large dams holding around 400,000 tonnes of hard pitch it wanted to get rid of. FFS mulled over the problem and came up with a novel solution. It discovered that Iscor[2] could make very profitable use of this material by injecting it directly into its steel smelters, thereby reducing the amount of electricity used.

---

[2] Our venerable old national steel producer Iscor has in fact been owned by the giant Indian steel corporation Mittal since 2005.

As the dams were situated near some mine ventilation shafts, the pitch could not be recovered mechanically, as the fine dust, being carcinogenic, would be a health hazard. So, the can-do lads at FFS welded up several kilometres of steam piping, put it into the pitch dam and created a pool of molten black sludge at the bottom, which they could then siphon off. The surface of the dam remained cold and semi-solid, and thus no fumes escaped.

Once the pitch had been vacuumed out, it was further heated and then filtered. At the peak of the operation, 22 tankers of 36 tonnes each were running to and from with the stuff every day. At the end of the operation, the last thin layer of pitch left in the bottom of the dam was simply, neatly and literally rolled up like a carpet and the "clean" dam handed back to Sasol.

"Over time we changed from being a company that just re-treaded oil waste to a technical innovations business," recalls Don Hunter, a bright young engineer who rose up through the ranks. When Hurter retired, Hunter became MD of what had up until that time been a strictly family business. "We would research, design, build, install and maintain new systems for clients, so they could use our products," Hunter says.

Each barrel of waste fossil fuel repurposed meant that one less barrel of crude oil needed to be dug out the ground, shipped to a refinery somewhere, then processed with all the attendant pollution and risk. Under Hunter's piloting, FFS was able to stabilise and control what had previously been a markedly insecure market situation.

When Sasol noticed how well the much smaller concern was doing financially, in 1987 the fuel-from-coal giant bought 49 per cent of the business (they later sold it, in 2005, as part of a BEE buy-in to FFS). By paying to take away the industry's waste, repurpose it and then sell it back to them at a good price, everyone made money, everyone was happy. Well, not everyone; there's just no pleasing some people.

At some point an awful "cat's pee" odour would drift over the Jacobs industrial park, Hunter recalls with a grin (by this time he'd been out of the game for some years). They were hardly the only heavy industry in the area, but they became the focus of local ire. A group called the Durban South Environmental Group (DSEG), led by environmental activist Desmond Desai, took up arms against FFS.

DSEG protested outside the FFS site and got lots of media coverage. One member of DSEG believed FFS was killing her, and she would come to the demonstrations wearing an old WWII gas mask (excepting, as someone pointed out, it no longer had its filter attached).

The only problem was, no one ever – still to this day – found out where the pong was actually coming from. "We were completely transparent," Hunter says. FFS did all the due diligence the health and safety authorities asked it to. It conducted its own investigations and made the data publicly available.

Still, no one could find the source of the feline stench. Desai continued his hate campaign against FFS but never engaged with the company. He refused to meet with managers when they invited him, but continued to vilify them in the media and at public meetings. One day, his group carried a coffin to the company premises with "Don Hunter" written on it. A great publicity stunt, excepting it wasn't true.

In retrospect, Hunter believes the stench might have been the result of two chemicals emanating from different sources interacting and forming a cloud. "There was a carpet factory up the road from us that used some terrible chemicals," he remembers. "But they never let us or anyone else in to see what they were up to. If we had also played closed shop, we could have saved ourselves a lot of trouble, and money.

"Desai was on one big, money-making ego trip. There's a problem when you have a complex issue but you are dealing with irrational people. And Desai ended up doing the environmental cause in the area substantial harm," Hunter believes.

Excepting, in the process of cleaning up its own business, FFS got even better at what it does. "There's money in dirty stuff," Hunter says. "We – the industry, and the nation as a whole – should stop focusing on how dirty it looks and find novel ways of dealing with it in more efficient, less polluting ways."

That's exactly what FFS has been doing for 40 years.[3]

---

[3] I conducted a series of interviews with the former managing director of FFS, DW Hunter, in January 2020 – it's a story I've been following for the past 20 years.

From GMO to OMG!
How Farming Evolved from Village Venture to Terminator Technology

From the time when wild pigs, sheep, potatoes, wheat, rice, bananas and many more animals and plants were being domesticated (what historians refer to as the First Agricultural Revolution) – well before anyone had ever thought to write their name on a piece of papyrus – no major advances were made in agriculture right up until Jethro Tull invented his seed drill around 1700.

That was the first of a number of mechanical innovations that helped usher in the Second (originally British) Agricultural Revolution. As for irrigation and the use of fertilisers, those were well established before any pharaoh thought to build a pyramid to demonstrate his immortality.

The introduction of seed drills, horse-drawn hoes and other farm engines through the 18th and 19th centuries had little effect on the food itself, but much more on the lives of the rural people. It ushered in the age of mass urbanisation and creation of middle (trading and crafting) classes around the world.

The use of seabird guano greatly facilitated the rise of intense farming of basic crops such as cotton, wheat, barley and clover (for fodder). It also initiated guano wars between the various industrialising nations, who all wanted to get hold of this "white gold" from islands off the South American and southern African coasts. During the American Civil War, Union and Confederate battleships betimes slugged it out in the southern Atlantic because not only is guano great for fertiliser, it was also a major constituent of gunpowder.[1]

Then came two world wars, and that changed everything rather swiftly. Wars will do that: Companies that for several years had turned to making aeroplanes, tanks, guns and gunpowder found that, after victory day, they had to find new ways of churning out saleable products overnight. Following fast on the Armistice of 1918, former arms companies started manufacturing and selling self-propelled, petrol-powered tractors, harvesters, furrowers and planters.

Hot on the heels of the Allied victory in 1945, makers of gunpowder flooded the world with synthetic fertilisers. A combination of factors – including mechanisation, lower food prices, increased wages and middle-class lifestyles – meant the business of farming got bigger, then much bigger.[2]

The counter-balance to this has been a decrease in the number of people actually doing the farming: In industrialised regions it has fallen between 25 and 40 per cent since the end of World War II. Over the same period, the number of individual farms in industrial farming regions has dropped an estimated 90 per cent. Increasingly, independent farmers have been earning steadily less, as vast-scale agriculture, combined with various systems of government subsidies, have forced them off the land.

IF YOU COULD CREDIT ONE person for the world's population reaching its current staggering 7 billion plus people, it would have to be Norman Borlaug.[3] In the early 19th century, English cleric and economist Thomas Malthus famously predicted that, all too soon, Earth would not be able to continue feeding its seemingly ever-growing population. Enter Nobel Prize-winning Borlaug, the main man behind the Green Revolution.

This American agronomist pioneered new strains of high-yielding crops, the use of hybridised seeds as well as the ubiquitous use of

---

[1] thebreakthrough.org/articles/remember-the-guano-wars; and en.wikipedia.org/wiki/CSS_Alabama%27s_South_African_Expeditionary_Raid

[2] In the first quarter of the 20th century, a typical middle-class family was estimated to spend between 30 and 40 per cent of its income on food. By the end of that century, the figure was closer to 10 per cent.

[3] https://en.wikipedia.org/wiki/Norman_Borlaug (accessed 8 September 2020)

synthetic fertilisers, herbicides and pesticides. It has been estimated that all this has managed to feed at least one billion extra people, but its knock-on effects have been catastrophic for nature.[4]

Take Roundup for example, a herbicide produced by the Monsanto Company (net income in 2018 US$2.3 billion), and probably the most controversial agricultural product ever made and sold, although not the most poisonous. That would include charming products such as Agent Orange, dioxin, DDT and PCBs, the very products that helped make this same company the giant (some would say giant monster) it is today. The company was founded as a chemical business in 1901. In the 1950s it became one of the faces of the future with various plastic products, including a modular "future home" that was exhibited at Disneyland.

In the 1970s it sold off much of its chemical and plastics interests and invested heavily in agri-business, buying up other smaller businesses that were pioneering seed biogenetics and associated products. It was in this year that a chemist working for Monsanto, John Franz, discovered that N-(Phosphonomethyl)glycine – otherwise glyphosate – was extremely effective as a broad-spectrum herbicide for controlling weeds in croplands. With that, Monsanto joined the Green Revolution by offering the glyphosate-based product Roundup to farmers. With its chemical background, the company also invested heavily in genetically modified (GM) crops.

Over the past few decades, there has been a growing distrust in GM foods, to the extent that some food stores will not stock them. "Fools!" reply the futurists: All our food has been GM since the ancients selectively bred their cattle for more meat, more milk or shorter horns, their horses for speed and endurance, and selected wild grasses for ever bigger and juicer seed heads.

While all that is true, that kind of GM has little in common with today's laboratory craftspeople, who tinker with genes and

---

[4] While Borlaug helped to feed around one billion people, at least six of the remaining 6.5 million can thank the food on their table for the German chemist Fritz Harber who, around 1900, worked out how to "fix" nitrogen – the most abundant but inert substance in our atmosphere, which is essential for all plant life, much like vitamin C to us. Using extreme heat and pressure, and adding the element osmium as a catalyst, he fixed it into NH_3 (ammonia) and thereby created the base material for all modern fertilisers.

chromosomes behind tightly closed doors. The apple William Tell shot off the head of his son was likely no sweeter than a typical carrot today.

Writing in the *Modern Farmer* in April 2014,[5] American journalist Nathanael Johnson explains that the GM argument is really a metaphysical one. On one side are those who believe that technological innovations are hurting us in many ways we know and as many that we don't. On the other side are those who believe technology will lead us bravely into a workable future. According to the latter argument, Monsanto just happens to be a "business as usual" bunch of people whose woeful PR has thrust them into the role of agri-tech villains.

One of the "Frankenfood" aspects of the GM business is that, while its products are extremely resilient against certain pests, and efficient in making quick profits, from what I have read they are also subject to patents and strict licensing that lock farmers in legally and practically. One needs to be very careful when signing a supply contract with Monsanto, their contracts are water tight.

The first shocker for unsuspecting farmers comes by way of "genetic use restriction technology": farmers find the seed they bought from the nice agri-salesperson is sterile, and the seeds they collected from the previous year's crop do not germinate. It's known in the trade as terminator technology.[6] So, instead of using profits from that crop to buy a new bakkie or whatever, they have to use it to buy more – sterile – seed. If they've signed a contract with one of the major seed producers, they find they're locked in for life.

Then said farmers also find they have signed themselves onto a treadmill of escalating prices, both for the seed as well as for the herbicides that have been especially concocted to work for – and only for – that particular strain of Roundup Ready crop. And that is before they – or we – even get to all the related environmental issues.

Monsanto's business model is built on the presumption of litigation. This came to world attention in the company's case against

---

[5] https://modernfarmer.com/2014/03/monsantos-good-bad-pr-problem/ (accessed 8 September 2020)
[6] cases.open.ubc.ca/monsanto-and-terminator-seeds/

Canadian farmer Percy Schmeiser, who was sued for illegally growing Monsanto-licensed canola. The farmer countered that the seed had blown onto his property. Although, and incredulously, Monsanto won the case, it was a Pyrrhic victory.

Crop seed, writes Johnson, is universally seen to be like air and water – universal gifts from God. They should not be subject to laws in the same way that computer software or some new tech innovation is. And all this is still only the veneer of problems that go deep down into the soil of modern agriculture.

In 2013 one of Monsanto's products, a Roundup-resistant strain of wheat, was found to have jumped from a lab in Illinois or Alabama and landed on a farm in Oregon. If you know your American geography, that's quite a leap. Monsanto insisted it was a once-off freak accident, but it stoked Frankenfood fears far and wide.

During the Vietnam War, Monsanto was one of the main producers of the massively destructive herbicide Agent Orange. In 1984 the company settled out of court with American veterans for harm caused.

In 2003, the past came back to bite its butt even harder. Throughout the 1960s and 1970s, Monsanto was a major producer of PCBs (polychlorinated biphenyls) that were used for electrical insulation, as coolants, in carbonless copy paper, aerosol can spray and heat-transfer fluids, mostly in fridges. When it was discovered PCBs were one of the main contributors to the depletion of the ozone layer (Earth's natural sunscreen), they were outlawed.

What the chemical giant did was simply dump what it had left into Dead Creek outside its plant in Sauget, Illinois.[7]

Later, a bunch of lawsuits were brought against the company, on behalf of 20,000 residents or former residents of Anniston. And subsequent investigations revealed it had been dumping mercury and PCB waste into streams around its plant in the Alabama town over a period of some 40 years. In 2003, the company agreed to pay US$700 million to settle the continuation claims.[8]

---

[7] beltmag.com/a-short-way-to-hell-in-sauget-illinois-poisons-mean-profit/
[8] The settlement was widely covered in the media, such as in this article

In 2018, the German pharmaceutical colossus Bayer bought Monsanto for US$66 billion. But, with that, it also bought a heap of legal trouble down the line.

Looking south to Argentina, we find just one of many controversies raging between Monsanto and local farmers. In 1966, that country's agricultural authority approved Roundup Ready soya beans; over the following decades, soya plantings increased from around seven to 20 or more million hectares. There was a corresponding decline in the production of milk, rice, maize, potatoes and lentil production for local markets, whereas soya production was being driven by investors for offshore markets.

Intensified, large-scale soya bean farming displaced as many as 150,000 small farmers in the South American country. This is just the kind of thing against which Alexander von Humboldt, the "man who invented nature" and the star of the first chapter of this book, railed, excepting now it was capitalism, not Spanish colonisation. In Buenos Aires Province, the effects have been most severely felt, with an eightfold increase in the use of agri-chemicals and a simultaneous rise in the incidence of various cancers among farmworkers. The main villain? GM soya and the broad-spectrum glyphosate Roundup on which it depends.

Monsanto has done all in its power to distance itself and its products from this link, helped in large part by the World Health Organization (WHO) that over decades insisted that it was "not sure" of any links between the weedkillers and sickness.[9] At the same time, medical researchers at the University of Buenos Aires had found an incontrovertible link between the two.

In the Al Jazeera TV documentary *Argentina's Bad Seeds*,[10] filmmakers Glenn Ellis and Guido Bilbao claim there is a revolving door through which Monsanto executives enter and leave the country's agriculture department "in a parlour game predicated on

---

from *The Guardian* https://www.theguardian.com/business/2003/aug/21/9 (accessed 27 August 2020)

[9] https://www.aljazeera.com/news/2016/05/monsanto-herbicide-glyphosate-cancer-160516122251162.html (accessed 20 September 2020)

[10] https://www.aljazeera.com/programmes/peopleandpower/2013/03/201331313434142322.html (accessed 27 August 2020)

making money". They also allege that resistance to Roundup had led to the increasing use of other – stronger, worse – herbicides such as 2.4D, which they claim has also been linked directly to rising rates of leukaemia, lymphoma, pancreatic and other cancers.

The United Kingdom, Brazil and China – each has its own beef with the big seed and poison company. In India, Monsanto's supposedly insect-resistant strains of cotton proved to be all but, which in turn led to widespread bankruptcy and suicides among small farmers, mainly in the Andhra Pradesh state.[11] The strain in question is codenamed Bt, which stands for *Bacillus thuringiensis*, a crystalline insecticidal protein that is manufactured by the seeds of this GM organism (GMO). In 2009 Monsanto discovered some insects had developed resistance to Bt, but the company did not bother to inform any of its customers.

With all of the reports and documentaries on Monsanto it seems that wherever there are seeds to be planted and money to be squeezed from the pockets of struggling farmers, there you will find Monsanto: and so, to South Africa. Early in the 21st century, the agricultural behemoth launched a supposedly drought-tolerant strain of maize (corn to Americans). It was branded DroughtGard, and the company tried to introduce it in the mealie fields of our beloved country (as well as several other southern and East African maize producers).

Monsanto claimed its drought-resistant maize gene (MON87460 x MON89034 x NK603) would increase yields for smallholders, especially in dry times. It even came with an endorsement of the éer-do-good Bill and Melinda Gates Foundation:[12] What could go wrong? Plenty, we have come to learn, and the Johannesburg-based African Centre for Biodiversity (ACB) said as much and laboured to disprove Monsanto's claims.

For 10 years, ACB put the suspect maize strain under the microscope, and by 2019 had found that not only did it not meet its maker's grandiose claims, in fact, in the words of the South

---

[11] www.permaculturenews.org/2010/01/26/farmer-suicides-and-bt-cotton-nightmare-unfolding-in-india/
[12] https://cagj.org/2019/05/failure-of-monsantos-drought-tolerant-maize-pushed-on-africa-confirmed-in-us/ (accessed 20 September 2020)

African Minister of Agriculture at the time, Thoko Didiza, it "did not provide yield protection in water-limited conditions and in some trials even showed lower yields than conventional maize".[13] To top it, claims that the maize clone was also insect-resistant proved to be scientifically unsubstantiated. Its release in South African was subsequently barred.[14]

According to ACB director Miriam Mayet, Monsanto's appeal was based only on differences in maize kernel characteristics: "The data exposes the twisting and manipulation of science by Monsanto to promote sales of their ineffective, reductionist GM products for complex environmental, political and socio-economic challenges, such as climate change and poverty."[15]

It is true that, initially, Green Revolution crops did show increased yields. However, here, there and just about everywhere, within a few decades things started to look less rosy on the food front. Firstly – GM crop yields levelled off (rice), and in some cases began to drop (maize). But the worst of it has been the emergence of plants ("weeds") and insects ("pests") that are resistant to the various pesticides and herbicides on which the GM crops are dependent. They have helped to breed the plant and animal equivalents of antibiotic-resistant superbugs.

Just like the tobacco and oil companies before it, Monsanto countered firstly by denial, then by paying denialists to sow discord and confusion. In this case, the principal villain was a medical doctor named Henry I. Miller – the same man who, in the 1990s, had promoted the products of the Phillip Morris tobacco company.[16]

Around the end of the millennium, reports started coming from the European Chemicals Agency (ECHA) and the WHO's International Agency for Research on Cancer (IARC), showing some possible health dangers posed by glyphosates to both human

---

[13] https://www.acbio.org.za/sites/default/files/documents/Minister%27s_final_decision_on_Monsanto_appeal.pdf (accessed 27 August 2020)
[14] *Ibid*
[15] https://www.acbio.org.za/en/resounding-no-monsantos-bogus-gm-drought-tolerant-maize (accessed 27 August 2020)
[16] http://en.wikipedia.org/wiki/Roundup_(herbicide); and http://usrtk.org/our-investigations/why-you-cant-trust-henry-miller-on-gmos/

and aquatic lifeforms. The dangers emerged during epidemiological studies (those dealing with the incidence, prevalence and control of diseases), animal studies as well as in vitro tests. In response, in 2015 Monsanto hired Miller to write a rebuttal in *Forbes* magazine.

However, after publication, the editors at *Forbes* received correspondence between the author and the chemical company that gave the lie to the piece. Notable was when Miller asked Monsanto to get the ball rolling by providing him with "a high-quality draft".[17] The magazine not only retracted the article, it also pulled his regular column.

Big farma is a very big business, to be sure, which is why Monsanto is so aggressive about it. Up-to-date agricultural statistics – like, how much poison we are using – are hard to come by in South Africa. A survey conducted in 2003 put the amount of poisons sprayed on croplands or otherwise dumped at about 10 million litres a year (value around R170 billion). The undiluted amount of agricultural poisons dispensed between then and now is around 200 million litres, enough to fill 15 Olympic size swimming pools.

Of the around 9,000 registered pesticides, herbicides and fungicides in South Africa (the continent's largest user) the majority are organophosphates. What they do is act like nerve agents, inhibiting neuromuscular enzymes in the insects they are sprayed on. They act similarly on other animals, including humans, but the amounts are usually not enough to do us much harm, unless you are a farmworker and they are sprayed around you. Or the wind blows them over your village. Or they accumulate up the food chain like DDT did in America in the 1950s.

It's like when you spray an insecticide in your home or garden to kill a fly, a mosquito or other gogga, you also kill bees, moths, butterflies and chameleons at the same time: Poison is poison. It's all connected. It's one of the reasons we see fewer and fewer of these small but vital creatures these days.

Around 90 per cent of South Africa's cotton and 85 per cent of our maize and soya plantings are Roundup-dependent GM varieties.

---

[17] https://www.nytimes.com/2017/08/01/business/monsantos-sway-over-research-is-seen-in-disclosed-emails.html (accessed 27 August 2020)

When you go to your local supermarket to buy your favourite sandwich loaf, regular brown, high grain or even the bread connoisseur's choice, you'll probably be getting a fair dose of Roundup. Even though there is no GM wheat grown in South Africa; however, almost all commercial brands of bread include soya bean, soya bean oil and/or soya lecithin. Gluten is the least of your worries.

In 2019 farmers across the Western Cape said, "enough Roundup already!" Around 200 women farmworkers from across the province marched on Parliament to demand the banning of 67 specific agricultural poisons to which they were exposed in their working lives. Many of them told of the lack of personal protective equipment, failure to be notified prior to spraying (including by aeroplane, which has significant wind-drift in this most blustery of provinces) and intimidation by farm owners when they voiced concerns.[18]

An organisation named Women on Farms Project reported that most workers come into direct contact with these poisons within an hour of fields being sprayed. The list of poisons include Roundup, Dursban, Paraquat and a host of others that have already been banned by the European Union, some as far back as 2007.

Israeli historian and best-selling author Yuval Harari writes that industrial farming has been one of the great crimes visited on Earth, not just on humans but even more so on the animals that are thus raised. Something like 90 per cent of all large animals alive today are domesticated, and most of them live in horrendous conditions. Writing in *The Guardian*, Harari says: "The march of humans is strewn with dead animals – sentient beings."[19]

When humans first reached Australia around 45,000 years ago, they quickly wiped out around 90 per cent of all the large animals there. It was the first time, but not the last, when humans had a direct hand in mass extinctions. Some 20,000 years later, when modern humans reached North America, they did the same.

---

[18] en.wikipedia.org/wiki/Roundup_(herbicide); and usrtk.org/our-investigations/why-you-cant-trust-henry-miller-on-gmos/

[19] Harari (b. 1976) is most famous for his books *Sapiens: A Brief History of Humankind*, *Homo Deus: A Brief History of Tomorrow* and *21 Lessons for the 21st Century*. He is a professor in the Department of History at the Hebrew University of Jerusalem.

Here are some statistics for people who like numbers: there are fewer than 20,000 lions living on the planet today, but there are about a billion farm pigs; there are fewer than 400,000 elephants, but there are 1.5 billion cows; there are around 50 million penguins, but something in the region of 20 billion chickens. Farm chickens in Europe outnumber all the wild birds there. Welcome to the new world; we hope you enjoy your stay.

Most of the deforestation and runaway fires in places like Brazil, Indonesia and Borneo are caused by farmers clearing land to plant maize and soya beans to feed cattle, or palms for oil used mainly in baked products. The rest will be used for pasture for cattle that will end up as hamburger patties. As the late great celebrity chef Anthony Bourdain put it: "There is a lot of great food in America, but the fast food is about as destructive and evil as it gets. It celebrates a mentality of sloth, convenience and a cheerful embrace of food we know is hurting us."[20]

It is becoming evident that some of the worst perpetrators of agricultural malpractice are plant farmers – soya, oil palms, almonds, in fact all large-scale plant production of monoculture that are forced to use artificial fertilisers and poisons because the natural balance of things has been broken.

The answer lies more in knowing how our food was produced than whether it is plant or animal. Growing GM maize and soya, or burning forests in Indonesia to clear land in order to plant oil palms, or almonds that depend on fossil fuel fertilisers and the heavy use of pesticides, is as bad as raising feed-lot cattle. Even most plant-based burgers and sausages still contain some bad products.

There is destructive farming, and there is regenerative farming. Mostly and simply they can be separated by the size of the operation. You should be wary of anything "mass". Remember, small is beautiful.

---

[20] www.nationalgeographic.com/adventure/adventure-blog/2011/08/03/anthony-bourdain/ (accessed 8 September 2020)

Farmers from A to Z
Small Farmers, Good Farmers and Bad Farmers

IT IS A SELF-EVIDENT TRUTH that you should never speak ill of a farmer while there is food on your table. And yet, driving past many dairy farms you'd want to stop your car, jump over the farm fence, knock on the farmer's door and pie them with a freshly laid cow pat, bellowing: "You should be ashamed of yourself; no self-respecting person, farmer or no, should ever treat animals like that!"

On a dairy farm in the Overberg I saw a field where cows were staggering, udders the size of Pilates balls forcing their legs out as if they had some grotesque deformity. Call them Farmer B, B for bully. Don't be like Farmer B, of whom I would not hesitate to speak ill.

Some distinguished people such as Jared Diamond[1] have said farming is the very worst thing humans have ever invented, given the effects it has had on nature over the past 10 millennia, and then on ourselves over the past 10 decades. Diamond says agriculture is a disaster from which we have never recovered.[2]

As has been noted in a previous chapter, agriculture began around 10,000 years ago. Ironically, while the human want for meat, milk and so on should ensure the wellbeing of the animals from which they are sourced, pretty much universally the complete opposite has become the reality. Publishing phenomenon Yuval Noah Harari reckons the

---

[1] Jared Diamond (b. 1937) is an American geographer, historian, anthropologist and author of popular science books.
[2] https://www.discovermagazine.com/planet-earth/the-worst-mistake-in-the-history-of-the-human-race (accessed 27 August 2020)

way farm animals die is one thing, but how they are forced to live is even worse. He says domesticated farm animals have paid the price in suffering for the astonishing triumph of human agriculture.

The root of the problem, he continues, is that domesticated animals still carry the genes and instincts of their wild ancestors, as well as their social and emotional needs. We force them to live in tiny cages, and their young are taken from mothers before they are weaned (often right from birth). The animals suffer greatly, yet they live on and multiply.

Even the ways they are fed are often unpleasant and – at worst – cruel. Take for example cattle: For some millions of years, bovines have developed a unique four-compartment stomach and associated bacteria suited for eating and ruminating grass and pasture herbs such as clover.

And yet, in most industrialised farming situations, they are fed maize and soya, sometimes worse, like repurposed offal and fishmeal. That way you don't need lots of expensive land to feed them. In fact, the only land you need is that to keep them enclosed, wallowing all day – often on cement floors – in their own urine and faeces. Or in deep, perpetual mud that causes their hooves to rot.

It does not have to be that way: rather than being like Farmer B, be like Farmer A. He's a real farmer, and a damn fine one. His actual name is Angus McIntosh. I became aware of this good farmer through various Facebook connections and when our family started buying his products wherever we could find them.

Visiting his blog, I learnt he grew up on a cattle ranch in the thornveld belt of KwaZulu-Natal. After studying accounting management at Stellenbosch, Angus met and married a local lass, and they had a couple of children. The dream took them to London and a job with one of the world's financial juggernauts. Things were going capital, as they say, until the masters of his universe told him his future with them lay to the east, Far East.

The couple started thinking of their children and the sunnier south, where Ms M's family owned an estate outside Stellenbosch. You might have heard of it, it's called Spier, a tract that had been worked to a state of agrarian fatigue since its founding in 1692.

The front of house at Spier is a well-known and typical show wine estate of the region, with complementary enterprises including a boutique hotel, bistro café, wine shop, art installations and a raptor rescue centre. But that is just the 150-hectare veneer of the 650-hectare whole. Sweeping up the slope to the east is a real mucking and trucking 250-hectare farm that has been turned into one of this country's premier organic producers (the remaining 200 or so hectares is a fynbos restoration zone).

To get to the working end of the operation you have to go via a side entrance on the Allandale Road, through a security boom and past a small school.

The open plan farm office provides a number of clues as to the nature of the operation. In the entranceway is a 40-photo organogram of all the farm workers (many of them Zulu-speaking people from the McIntosh family's old farm in the Weenen district, which was expropriated in a land claim), without any obvious "boss" pyramid. Large display boards show distressing images of industrial farming in places such as the United States and Brazil. They are like "before" photos in diet adverts, and they are ugly as sin – caged chickens in warehouses, cattle in feedlots, forests burning.

As pre-arranged, Angus arrives, in shorts, barefoot and an irradiance of energy. He fires off instructions to various people in the office, then signals to me that we have just a one-hour window in his day, so best to get going. First thing he puts straight is: "We practise regenerative agriculture here. Vegetarianism is a personal choice, but it will not save the planet." If you take out the animals from farming, he explains in some detail, there is a net loss of nutrients in the system. Clearly, this kind of farming is more than just a matter of planting mealies and praying to the powers above (for rain).

Returning home from England in 2004, and desperate to go off the grid, Angus began building a clay home for the family and, following that, clay homes for the farm workers when he became manager of the estate.

"I saw how ecosystems were – are. So stuffed up, not just overseas but also here as well," he says.

Reading about the Salatin family of Virginia, United States, in the book *The Omnivore's Dilemma* by Michael Pollan,[3] was the big game changer. In 1961, patriarch William Salatin bought a worn-out, abused and eroded piece of land in the Shenandoah Valley and began the long process of restoring its health. Three generations later, and after much tree planting, composting, rotational grazing and other good farming practices, today Polyface Farm is considered America's showcase regenerative farm. So, there was Angus on a showpiece farm, but he could see how things could be done much better.

With local regenerative farming guru Christo Kok as his mentor, Angus began driving Spier down the road of organic farming: no chemical fertilisers, no herbicides containing glyphosates (the dreaded Roundup), no animal antibiotics, no GMOs. Instead, he would be predominantly a pasture protagonist (with some orchards and vineyards). And thus was Farmer Angus born, the Farmer A of our agrarian parable.

On our hour-long excursion I learn that Farmer Angus is one of only two producers of genuine grass pasture-reared beef in the Western Cape. The farm's meat contains no added nitrates, nitrites, gluten or anything else. Furthermore, all his farm animals – pigs, chickens, cattle – are guaranteed not to be fed on anything that contains glyphosates. He is particularly proud of a new innovation, the only beef burgers – in the world, he claims – that come in compostable packaging.

The 4x4 crests a ridge from where we look down on the entire farm, and the Eerste River valley. This brings us to one of the farm's hallmark installations: the 20 "eggmobiles": laying enclosures that can be wheeled around by just two people, albeit two robust Zulu men in this case. Each eggmobile shelters 325 laying hens, which are free to lay up inside, underneath in the shade, or to wander about scratching in the ground as fowls were meant to do.

The mobiles are one of the innovations he borrowed from Polyface. They are moved daily, and portable fencing follows. Each laying basket is filled with diatomaceous earth in which the hens dust bath to get rid of skin parasites. Simple, like so many things, when

---

[3] *The Omnivore's Dilemma: A Natural History of Four Meals* by Michael Pollan (Penguin, New York, 2006).

you have thought the whole thing through.

The real magic of his system is the feathers, Angus tells me. Feathers that have been shed are collected and used as mulch in the orchards and vineyards. "We're on Table Mountain sandstone soil here," he explains. "It's very acidic and nutrient-poor, but every time we do soil tests, the quality has gone up."

The herd of around 300 cattle is moved onto new pasture two or three times a day on a three-month rotational cycle, depending on season and conditions. The "massaging" of hooves, with manure being added continuously, keeps the grassland in a healthy state. "You have to experiment. There are no hard rules. Each farm and each farmer is different."

As we approach the pigs skoffelling in the dirt on the valley floor, he chuckles as he tells us about the SPCA lady who had recently accosted him for feeding the porkers poisonous plants: someone driving past had seen them feeding on Paterson's curse, a purple-flowered invasive weed (in South Africa, at least) that goes by the botanical name of *Echium plantagineum*. While it is toxic to some grazing animals, especially horses, the pigs – being omnivores like us – were tucking in with relish.

"They cleared this field in three days," Angus smiles, scanning what looks like a recently ploughed paddock. Free-ranging pigs are nature's tractors, and on Spier they are used to clear and turn over fallow fields that have lain weed-choked for years. "We plant summer squashes behind them as they go." The pigs are moved to a new patch every three days.

As my time with Angus is running out, I drop my 50-million-dollar question: How does he compete commercially against all the other highly industrialised, mass-production farmers?

"Simple," he shoots back. "No, or fewer, middlemen." He sells only through select local and regional outlets as well as off the farm. "Every time you make a food purchase, you are choosing a farmer, but are they destructive or regenerative?" he leaves that dangling.

We have vast swathes of monoculture in this country, Angus expounds – what he calls "lazy" farming. There is no place for laziness in his worldview. "Look at all the sugarcane in KwaZulu-Natal,

Mpumalanga, Limpopo. It is a highly polluting form of farming, and we don't even need it."

While piling me with boxes of his free-range eggs, Angus tells me the name he has given to his farming operation: Ezibusisweni. He explains it's a Zulu word meaning "place of blessings", but to me it sounds more like the noise of busy bees working. Farmer A is many things, but indolent is not one of them.

Driving home, I wonder: If you can farm regeneratively on 400 hectares, then why not 4,000, or 40,000? My thoughts drift back to an article I read in some inflight magazine about how Jody Scheckter had become Britain's leading, possibly largest, organic farmer. Not something you'd expect from Sideways Scheckter, the reckless racing kid from East London in his white-and-blue-striped Renault R8 Gordini (number 139).

After Scheckter won the 1979 F1 world championship in a Ferrari, he lost his racing mojo and retired at the end of the following year. Next he popped up in the United States where, with other family members, he was involved in developing a firearm simulator training system for military and law enforcement agencies, without having to fire actual bullets. After selling that business he moved to the United Kingdom where he bought a farm, Laverstoke Park, in Hampshire, south of London.

He never meant to become Britain's favourite organic and biodynamic farmer, but he's always been a fast learner. The thing is, when you are running a working farm thousands of hectares in extent, you need to feed more than just your own family, or most of what you produce goes to waste. Being the competitive person he is, Scheckter was probably determined to follow the very best farming practices – and that led naturally down a regenerative road. In doing so, he has demonstrated that you can scale up good farming practices.

"We are creating the most ideal, natural, healthy environment that will enable our animals and crops to thrive. We follow nature, but use the latest and best scientific research, techniques and equipment," says Farmer J on the Laverstoke Park website.[4] Laverstoke produces beef, dairy, pork and poultry products, as well as a range of beers, and more

---

[4] https://www.laverstokepark.co.uk/about-us/ (accessed 28 August 2020)

recently sparkling wine, with homegrown Pinot Noir, Chardonnay and Meunier varietals, using strict biodynamic methods.

The website continues: "This begins with the soil – soil is 90% of farming – there are more living organisms in a handful of good soil than people on earth! By enhancing the healthy bacteria and fungi in our soils, this aids plants to absorb the nutrients effectively and that is what gets the nutrients from the soil to the plant."

But farming alone was never going to be enough for a man who tinkered with tappets and timing chains in his youth. Laverstoke Park Laboratories (licensed by the Soil Foodweb Foundation) offers soil biology advice as well as chemical analysis to anyone who asks.

Laverstoke's dairy products – including cheeses, unctuous ice cream and milkshakes – are to die for, courtesy of the farm's water buffalo dairy herd. Laverstoke is a long way from Watkins Glen in the United States (where Jody drove his first F1 race in 1972, in a McLaren), but the objective is not much different: to be the very best you can in everything you do.

Good and well, but what has Hampshire got to do with Harrismith? Farmer D is what, which here stands for Danie Slabbert. Slabbert grew up on a typical eastern Free State plaas, where they grow stuff like mealies and wheat and sunflowers and soya beans and potatoes. It was considered quite alternative back then to grow the radishes that are now a popular rotational crop of the region.

What made Slabbert a bit different from the other plaasjapies of the district was that his folks sent him off to actually study agriculture, in far-off Saasveld, near George. That kind of thing will get a young man thinking, for good or for bad. Many people in the region thought it was bad, possibly even blasphemous when he came back and started talking about new-fangled methods of farming – like not spraying.

How the donder do you get rid of leafhoppers and thrips and stalk borer and all the other plagues that have been sent to punish farmers? Not to mention white grubs and rootworms and ground beetles that are clearly the Devil's instruments. Outside church, those who have attended will invariably look to the sky, for signs of rain or salvation. Farming here is like the lottery, they'll all agree: It's such a gamble, it's a wonder the Good Lord allows it at all.

But upon his return from Saasveld, Farmer D also offered some more comforting thoughts, like that farming was a calling, and that farmers had been put on the land by God. But then he went and spoiled it again by using awkward words like sustainability. One can only imagine what the rest of the community thought when he and a couple of other kêrels in the district got together to discuss these progressive ideas, and they went on to form the Riemland Study Group to sow them further and wider.[5]

Agriculture has broken down, Slabbert says, what with the use of chemicals and large machinery. "We've actually weakened the land without even knowing it. What we want to do, is bring it back to its natural state," he says.[6] He uses other modern words like regenerative, and natural and unconventional. It seems to have struck a chord with at least some other farmers in the area, because in time, the study group expanded into an agricultural research group whose central tenet is… wait for it… to not plough.

With ploughing, you turn over the topsoil and release the dormant nutrients for immediate take-up by new crops. That leads to a massive flourishing of new plantings. But every time you plough, things cascade: the topsoil dries out a bit more, and the bacteria and other soil enrichers including earthworms decrease. Then crop harvests begin to decline, so initially you add manure from your livestock.

Year by year that is not enough for intensely planted monocultures such as mealies, wheat or sunflowers, soya or potatoes. So, you have to add artificial fertilisers to maintain harvest levels. But by then you find diseases have infiltrated the weakened bio-organism that is your farm, and you begin to use poisons, herbicides, to clear the weeds, and pesticides to kill the bugs (bad as well as good ones). That rot sets in, and it escalates with each season of ploughing and spraying.

Most basically what the Riemland group is trying to advocate is to not disturb the land and to bring life back into the soil. Admittedly

---

[5] From "How a Free State farmer brought his dying land back from the brink" by Dawn Noemdoe, 22 March 2019 https://www.foodformzansi.co.za/how-a-free-state-farmer-brought-his-dying-land-back-from-the-brink/ (accessed 28 August 2020)

[6] *Ibid*

getting these methods accepted – to plant without ploughing, and to bring back stock onto what were previously monocultural croplands – has been challenging, to say the least. Not least of it being the short-term sacrifices (lower yields) that have to be endured for sometimes several harvests before the land bounces back, and yields start to increase.

"We want to bring back livestock. We want biology and organisms to live and work together. We want to use less chemicals and rebuild the land, so that it retains moisture and structure," says a man who has gone down a hard road to reach a better place. After a journey of 11 years, he can now say with confidence that: "My plants survive with less rain and my land's nutritional status has improved immensely."[7] Not only cereal crops but also potatoes and soya beans – and radishes.

If a farmer encounters a plague of bollworm in his soya plantation, the first instinct is to spray chemicals. But that is treating only the symptom: You kill the bollworms but also their predators. That way the symptom (plague) will just return later, and maybe worse.

With the return of earthworms, nature's own composters, he has seen the return of other creatures that have been absent from the area for decades, like owls (26 in one gathering) and blue cranes. What that means is that the soil has come back to life and with it the pyramid of life from microbes and mice to insects and birds.

"I'm trying to prove that there is actually enough space for everybody in South Africa," concludes Farmer D.[8]

THERE IS A LOOSE GROUPING of growers who, for the sake of the story, will be known collectively as Farmers M, who are helping to feed poor people one tiny food lot at a time. Think of the entrepreneurs who sell small piles of fruit and veg at stations and taxi ranks around the country: Where do you think they source their produce? From the likes of Farmers M is where.

What if you don't own a farm the size of a small English county, or a tractor, or even a wheelbarrow – what then? You make do with

---

[7] https://www.foodformzansi.co.za/how-a-free-state-farmer-brought-his-dying-land-back-from-the-brink/ (accessed 28 August 2020)
[8] *Ibid*

whatever you have. In places as far flung as Cape Town, Johannesburg and Bushbuckridge are small farmers, many of them women, who have taken to doing it for themselves, their families and communities. The names of the ones I found for this chapter all just happened to all start with an M.

Another thing they had in common was that they had each been upskilled by food security organisation Food & Trees for Africa, with a little financial foot-up from Shoprite.

Mama Mickey Linda lives in the vast shackland in the Cape known as "new home" (Khayelitsha). She started Yiza Ekhaya,[9] her own soup kitchen, to feed the needy in her neighbourhood with only her government grant money and what was in her kitchen.

On the first day of opening, she filled 40 hungry mouths and so, calling upon the prevailing ubuntu spirit, other eager helpers joined to help her run her soup kitchen and work the vegetable garden that she had begun to make the soup.

Meryline Ntimande has assembled a team of some 40 helpers to tend a food garden in a place called Mabarhule near Hazyview in rural Mpumalanga Province. Her Tiyimiseleni Home-Based Care scheme feeds around 1,000 people each day.[10] The turning point for her was learning from Food & Trees how to compost and mulch her previously poor soil.

Margaret Thaba and Miriam Malunga from Alexandra in Johannesburg tell similar stories, their Modimo O Teng (God is Here) project having grown from a small garden on an empty lot to a small farm tended by many.[11] Even in the city's leafier suburbs, there are people who have turned pavements in the City of Gold into food gardens for the passing needy.

And not only pavements but also gardens of residential estates, apartment rooftops, parks – and even a repurposed old commercial

---

[9] https://yiza.greenhome.co.za/ (accessed 28 August 2020)
[10] https://www.shopriteholdings.co.za/articles/Newsroom/2018/Meryline-proves-that-one-woman-can-make-a-huge-difference.html (accessed 28 August 2020)
[11] https://www.shopriteholdings.co.za/articles/Newsroom/2018/Good-soil-and-great-partnerships-are-the-key-to-improving-lives-at-Modimo.html (accessed 28 August 2020)

block called the Victoria Yards in the decaying inner-city Johannesburg precinct of Lorentzville (my farmers Z).

From the poorest to the richest neighbourhoods, the spirit of ubuntu, grace, still falls like gentle rain – or manna – from heaven. All you need is some soil, a handful of seeds, a can of water and a sprinkling of generosity.

A REGENERATIVE FOOD FARM CAN be as small as an old coffee tin or as big as your dream.

Busy Bees
Dying Bees and Busy-as-Bees Honey Embezzlers

AT A MEETING OF THE Royal Geographical Society in London in 2019, Earthwatch Institute stated that 'bees are the most important species on the planet'. This statement is timeous given that they are facing extinction in much of their ranges. An estimated 70 per cent of all the food we eat is dependent on insect, mainly bee, pollination, directly or indirectly: clover, sheep, lamb casserole, just for example.

Not only that, but without the fruits and flowers bees help to propagate, many animals, both wild and domesticated, would also lose their food sources. There is a number of known and supposed reasons for the disappearance of the bees, the most obvious being poisons sprayed on flowering crops and the loss of natural habitat and forage plants.[1]

But there are also some less obvious ones, including stress and the mesh of electromagnetic waves that envelopes the planet. Swiss Federal Institute of Technology tests show that when bees are subjected to mobile-phone waves, they emit piping or "mayday" kinds of distress signals that tell their colonies to evacuate.[2]

---

[1] Much of the technical information in this chapter is taken from South Africa's standard beekeepers manual, commonly referred to as "the blue book" – *Beekeeping in South Africa, Third Edition, Revised: Plant Protection Research Institute Handbook No. 14,* Department of Agricultural Technical Services/Agricultural Research Council, Pretoria, 2001.

[2] www.researchgate.net/publication/225679194_Mobile_phone-induced_honeybee_worker_piping (accessed 10 September 2020)

There are around 20,000 bee species in the world, some 3,000 of which can be found in Africa.[3] Nine species of honey bee occur worldwide but only one, *Apis mellifera*, the common honey bee, gives us the golden liquid we so love. The name means "honey-bearing bee" and they are united by the common traits of living in social colonies with an all-controlling queen, producing wax and, of course, honey.

For the record, there are 24 races (sub-species by some definitions) of *A mellifera* of which 10 occur in Africa. In South Africa, we have two types of honey bee, the Cape honey bee, *A m capensis* (which is almost entirely restricted to the fynbos zone or Cape Floristic Region), and *A m scutellata*, the African lowland honey bee. It is the latter, sent from a hive in Pretoria to a research facility in Brazil, that was released in South America and, when it reached North America, was labelled the "killer African bee".

We learn in school that bees dance, but it's far more complicated than just a wiggle here, a jiggle there, and a bit of buzzing around to the music of the spheres – although that is part of it. It was an Austrian ethologist (that's someone who studies animal behaviour) who discovered the true politics of bee dancing.

When World War I exploded across Europe, Karl von Frisch packed up the books and instruments at his laboratory in Munich, Germany, and absconded to his Austrian country house in Brunnwinkl. Walking in the meadows there, he began to notice the busy-ness of the bees and became fascinated. It was already known that bees danced, but Von Frisch wanted to learn more.

He dabbed worker bees with coloured dots and then followed them out foraging and back again to their hives. (Farmed bees live in hives, wild ones in nests.) By astute observation he first noted that they did the dance of the honey hoard: The first ones out of the hive when the day warmed up were scouts that, when they returned from a sortie, would indicate to the awaiting swarm the way to the best forage, where they would find nectar and pollen, the distance and even the strength of the source.

---

[3] Some references put the number at seven, others at eight, but nine is the current generally accepted one. Taxonomy (the defining of species) is a constantly evolving discipline.

For a long time, Von Frisch's claims were dismissed: How could creatures with exoskeletons and brains the size of pinheads share anything but basic instinctual information? The Austrian found that when returning foragers did their honey dance, they used the angle of the sun to the vertical plane (usually the side of a nest or hive, or other substantial object like a tree or a wall) in order to indicate direction. Even when the sun was not visible, they used either polarised light waves or gravity as their plumblines. The length of their line dance and its vigour indicated first distance and next quantity.

We now know honey bees also do various dances for different functions inside the hive, each one connected to a specific caste of worker bee (as distinct from the ones that involve the queen and her duties). All the many functions are performed by the female worker bees. The male drones are there for one and only one purpose: sex with a queen.

The "washing, ironing, cooking, cleaning and nursing" of worker bees consists of tasks including brood (wax cell) cleaning, removing dead bees (undertakers), ripening nectar, packing pollen, nursing brood cells (larvae and pupae), guarding, foraging (field bees) and ventilating and warming the hive or nest. The interior must be kept at a constant 33 to 36 degrees Celsius. In the heat of summer, this is achieved by wing fanning and water evaporation, while through cold winters the swarm clusters ever tighter and buzzes to generate body heat.

Worker bees live for about three weeks, and each job corresponds to an age group. Immature stay-at-home workers do most of the hive tasks. After about 10 days, they begin to take short navigational flights, and around two days later they begin collecting pollen, nectar for honey and resin for propolis. Pollen, or bee bread, is the adult bee's main source of food; nectar is collected in a bee's honey stomach and, on returning to the hive, it is passed from bee to bee for about half an hour all the while getting concentrated.

Propolis is a tacky, black substance, vital to the wellbeing of a hive or nest. It is made from sap gathered from flower buds and gum from trees such as pine and eucalyptus. It is so important for sealing the colony's home, when natural resin is not available, bees have been known to gather tar from roads and even paint from old tins (which is why commercial beekeepers are advised to keep their hives well away from those sources).

All that working and then dancing is fascinating enough, but it begins to look pretty mundane when you notice what the drones and queens are getting up to. These are subjects that would delight Dr Tatiana, the "Dear Abby" of the animal world.[4]

All the males do is mate and then drone on all day long until the workers have had enough of them and chase them out of the colony when food runs short. Only a small percentage of a hive's population consists of drones: there might be as few as 20 in an entire colony of 30,000 or 40,000 bees. When they sense the time is right, the drones fly off to congregation areas, mid-air mating zones that are located a substantial distance from any hive in order to ensure that drones from as many other colonies as possible all gather in the same place. This ensures diversity of each hive's genetic make-up.

Several thousand drones might gather in a congregating zone. No one has yet figured out how they know when or where to gather, but some of these areas have been recorded to persist for more than 100 years. When you see a swarm of bees hovering somewhere, it is most likely a gang of drones waiting for a queen to arrive. They cannot sting; only the females can, with their adapted ovipositors, but such swarms frequently send humans scattering.

When a queen is ready, usually one to two weeks after birth, she will leave the hive and search out a congregation area. As the virgin queen enters the mating territory she emits a pheromone (9-hydroxy-decenoic acid), which gets the drones all revved up. They will form "comets" consisting of several hundred bees each and go zigzagging across the queen's scent trail until they locate her.

When about a metre from her, the drones get their first visual contact and then they mob her, jostling to mate with her. The act of ejaculation is achieved by an air sac that compresses and expels a drone's genitalia along with his sperm. Once mated, the drone falls to the ground, paralysed.

---

[4] Olivia Judson, author of the hilarious but scientifically accurate book *Dr Tatiana's Sex Advice to All Creation: The Definitive Guide to the Evolutionary Biology of Sex* (Vintage, London, 2002)

The next drone has to scrape off the previous one's sperm sac before it can make its own contribution to the royal reproductive chamber.

A queen will mate with as many as 20 drones on each flight; over a week or so she will make between 24 and 44 mating outings. When she returns to the hive with the genitals of the last drone to mate with her still protruding from her oviducts, royal chamber maids remove it and clean her up.

Only about 10 per cent of all the spermatozoa released into her oviducts reaches the queen's spermatheca, or sperm storage organ. Even so, that will be more than enough for her to fertilise and lay up to 2,500 eggs a day for the rest of the approximately four years of her reign. A queen might die of natural causes, be hawked by a predator while out on a mating flight, leave with a break-away swarm if the colony gets too big for its quarters or flee if their home is overrun with enemies – of which there are many.

If a queen departs with just part of the swarm, workers left behind will create several queen cells, which are much bigger than the other brood cells and resemble peanut shells. They will feed the larvae exclusively on royal jelly, which turns an ordinary larva into a sovereign one. During this time, internecine warfare breaks out in the hive: It takes 15 or 16 days for a queen to fully develop, and during that period there will be a massive die-off.

The first queen to hatch is immediately coronated, and she will break open the unhatched royal cells and sting the emergent princesses repeatedly until death. Should any of the other queen cells hatch, workers will cluster on top of it to restrain the pretender while the chosen one stings it to death (queens do not die after stinging): A colony sanctions only one ruler. It is certainly not all dancing and singing inside a beehive.

Some fun facts about bees (specifically South African honey bees):
- It takes about 1,200 bees to gather nectar from about 2 million flowers to make one 500-gram jar of honey.
- Bees have no intention, and indeed no knowledge, of pollinating flowers: it's a free service they render in exchange for the flower giving them pollen and nectar. Isn't nature terrific?
- Bees beat their wings around 200 times a second and can fly

up to 25 kilometres an hour. Their normal range for seeking out forage seldom exceeds about 5 kilometres. However, in desperate times, they have been recorded venturing more than double that distance in search of sustenance or water.

- Foraging bees eat honey before setting off and will consume some nectar on a long return flight.
- A bee will visit between 50 and 100 flowers on each foraging trip. They might make as many as 12 trips a day, so visit and pollinate up to 12,000 blooms.
- The average worker bee will fly about 600 kilometres in her life in order to contribute about half a teaspoon of honey to the combs.
- 20 kilograms of honey converts into 1 kilogram of wax.

BIOGEOGRAPHY DID NOT BESTOW THE patronage of honey bees on the Americas or Australia: The honey industries there use mainly imported European honey bees, most common of which is the relatively docile Italian honey bee *A m ligustica*. They were first introduced to the east coast of North America by English settlers in 1622, and 231 years later (1853) were first noted on the west coast. The African honey bee took much less time to get there.

In a project to cross-breed bees, Brazil imported a number of races in 1956, including the African "killer" variety *A m scutellata*. A year later they escaped their quarantine (some say they were let out) and promptly headed north. They were first recorded in Texas in 1990 and fear flowed ahead of them like a panic wave. They were bigger, more robust and aggressive than any other honey bees they encountered and so easily outcompeted them.

By 1993 they had reached Arizona, New Mexico, and southern California in 1995, and by 1998 they reached Nevada. So docile were the established races compared to the new arrivals, up to then many beekeepers hardly bothered with protective clothing when handling hives: you would not dare that with the rapacious African swarmers. Not long after this, many US states began reporting massive swarm losses of commercial hives, used there mainly for agricultural pollination.

A new phenomenon arose, known as colony collapse disorder, or CCD. Losses of 30 to 50 per cent a year, in some cases as high as 90 per cent, were reported. No one really knew why. Around 75 per cent of all commercial crops – including soya, canola, clover, almost all fruits and nuts, notably the almonds of California – depend on commercial pollination. First the African bees got the blame. However, as more information accumulated, other causes became apparent; none more so than the potent neonicotinoid pesticides that are much favoured by American farmers.[5]

In the early 2000s, widespread CCD was noted across Europe, the worst-hit area being Germany's highly developed Rhine Valley. Hive mortalities were so high, in some places special bins were placed along autobahns for beekeepers to dump old empty hives. It was found that one specific class of insecticides, neonicotinoids, was the worst villain, and one specifically, Clothianidin, that can be 10,000 more powerful than DDT. It took a decade for that product and other similar neonicotinoids, which disarranges the internal wiring of insects, to be banned by the European Union.[6]

Increasing outbreaks of corn rootworm had spurred the use of ever more potent pesticides. In 2013, German beekeepers reckoned their pollinating business was worth about €15 billion a year. But German crop farmers were not impressed, as was pointed out in a *National Geographic* article, as the ban on the poisons would cost them around €880 billion in lost yields.[7]

Since then, another topic has come into bee colony collapse discussions: stress. People with AIDS who die do so of pneumonia, tuberculosis or some other bug that hits them when their immunity systems are compromised. Likewise, diseases are the tipping point of stressed bee populations. Starting in the 1980s, a widespread host of parasites – including varroa mite, foulbrood, wax moths, tracheal parasites chief among them – seemed to get the upper hand of honey

---

[5] Neonicotinoids are widely used pesticides implicated in the decline of bees. "A neonicotinoid pesticide impairs foraging", https://www.nature.com/articles/s41598-019-39701-5#:~:text=Neonicotinoids%20are%20widely%2Dused%20pesticides,ability%20to%20learn%20floral%20associations, 18 March 2019

[6] https://www.nationalgeographic.com/news/2013/5/130510-honeybee-bee-science-european-union-pesticides-colony-collapse-epa-science/ (accessed 20 September 2020)

[7] *Ibid*

bee colonies. What caused these plagues no one could say exactly.

The United States is the second largest farmer of commercial bees, after China. There is a sizeable community of itinerant bee farmers in the United States, some of whom don't have homes but live in their trucks while transporting hives state to state, farm to farm, as different crops come into bloom. The bee truckers might think it's okay, but bees don't like it much.

It takes a substantial part of their short lives to fix their location, then learn how to navigate out from home and back, in their search for forage and water. Moving them continuously drives up their stress levels and, when combined with the other factors, such as coming into contact with agricultural sprays, their immune systems become compromised and in sneak the baddies.

Many of the same diseases and pests have an impact on South African bees, some of which have been imported via improperly quarantined bee products, but nothing like the extent of America's. This is mainly due to the vigorous nature of our bees. Added to this is the fact that an estimated 80 per cent or more of our bees are still wild.

Along with that of songbirds, frogs, large predators and some other key species (bio-indicators), the health of bees tells us a lot about the health of our environment. If the bees go, we all go, goes the modern environmental dictum.

In much of the rest of the world, the chief concern is about commercial honey bee dynamics, but in our neck of the meadow friends of bees are much more concerned about the effects of commercial beekeeping on wild bee populations.

The threats to South African beekeepers are a bit more prosaic: vandalism, breaking of hives to steal the honey, honey badgers and veld fires. Prosaic or not, honey theft in some areas is driving bee farmers to desperation because the criminals – including those shaggy black-and-white marauders – invariably smash the hives to get at the sweet stuff inside

All the human misdeeds aside, one of the biggest natural enemies of our local bees is the large hive beetle. The female *Oplostomus fuligineus* (a type of flower chafer, 20 to 23 millimetres long) must feed on bee brood in order to lay her own eggs so will devour brood, pollen or honey cells indiscriminately.

In a healthy hive, should one of these beetles breach the outer defences, the workers inside will pack around it and cover it with propolis. Imagine a horde of Lilliputians covering you with wet fibreglass resin that begins to set in mere minutes. The small hive beetle (*Aethina tumida*) is much more insidious than the large variety. They can hide in tiny crevices and, left unchecked, an infestation invariably leads to colony collapse. Their voracious 10- to 11-millimetre-long larvae feed insatiably on honey cells.

In South Africa we might not starve if our pollinating bees disappeared, but imagine life without most of our deciduous and citrus fruits, cherries and most berries, onions and peppers, no beans or peas in the pantry, nor most kinds of nuts. About 100 of the plant foods we eat depend on insect pollination, from almonds and broccoli to youngberries and zucchini.

Farmers in general seem to have learnt how to spray judiciously, says John Moodie, one of the country's biggest beekeepers. "But macadamia farmers are still a bunch of cowboys when it comes to pesticides," says Moodie,[8] who also runs beekeeping courses on his farm Honeywood in the Grootvadersbos foothills.

Another problem for bees, he says, is the indiscriminate spraying by unenlightened farmers in places like the Langkloof, a major deciduous fruit-growing region between Oudtshoorn and Port Elizabeth, where Moodie places hives every year. "They will spray for fruit flies at the same time the orchards come into blossom without connecting the dots."

Which is not to say all beekeepers are among the most ardent of conservationists. For many of them, bees are like sheep, or chickens, or nuts – just collections of organisms that are a means (making honey or pollinating food crops) to an end (making money).

Take KwaZulu-Natal Midlands honey farmer Roland Kennard who, when caught in the act of selling honey diluted with sugar water, told *Carte Blanche* investigative reporters in July 2018 he liked to stretch "the boundaries of creativity".[9] He was one of five farmers in that area who were linked to honey fraud.

---

[8] John Moodie pers comm (during beekeeping course)
[9] As per a *Carte Blanche* episode that aired Sunday 29 July 2018

Unlike most other countries with significant bee businesses, the honey market in South Africa is substantially larger than the pollinating one. Some 3,000 tonnes of honey is consumed here each year, generating around R3 billion.[10] That gives lots of wiggle space for creativity.

It is thought that as much as 75 per cent of stuff sold as honey worldwide is concocted sugar water, most of it coming out of China.[11] The amount reaching South Africa annually is thought to be about 2,000 tonnes (two-thirds of the national consumption). Counterfeit honey often has pollen added to give it a deceptive chemical masking.

Even some of the world's most expensive honey brands have been rumoured to contain less – or more – than pure honey. (Beekeepers in the Cape contend their fynbos honey is at least as pure as the famous New Zealand Manuka brand that sells at around R650 for a 500-gram jar – at a fraction of the cost.) Imports aside, small-scale beekeepers I know believe that a substantial amount of all the honey sold in South Africa is made by bees fed on a diet of sugar water.

Another contentious issue is those flowering foreigners, Australian eucalyptus or gum trees, that beekeepers love but others loath. Conservationists deplore them because they are such ravenous consumers of water. Beekeepers, on the other hand, rely heavily on gums to provide nectar and pollen when the local blooms have all fallen. There are some 85 species of gum tree in South Africa, and if you planted them judiciously, they could sustain a bee business year-round.

CAPE TOWN BEE CONSERVATIONISTS Jenny Cullinan and Karin Sternberg have been observing wild colonies for several years. It took them about three months to get their eyes tuned in to "seeing" bees, and then they saw them everywhere. They say they watch bees like other people watch birds. Given that there is about the same number of wild bee as bird species in South Africa, it's not an altogether offbeat choice – other than the challenging size of the former.

The pair has identified more than 90 wild nests on the Cape Peninsula, with an average colony size of around 16,000 bees. They

---

[10] https://www.indexbox.io/blog/honey-market-in-south-africa-to-reach-3-5k-tonnes-by-2025/ (accessed 20 September 2020)
[11] *Beekeeping in South Africa,* see footnote 1 in this chapter.

believe between 85 and 90 per cent of honey bees in South Africa live in the wild where they are healthy and presumably happy. The biggest threat to them is being captured, put into hives and farmed to death.

In the light of all the above and what we know of our own species, it was going to be only a matter of time before people started playing around with bee genetics. Scientists in both Germany and Japan say they are well on their way to creating an insecticide-resistant "superbee" using Crispr-Cas 9 genetic editing technology.[12]

The German scientists insist they are not connected to "big farma" and that they are just researching bee behaviour. Unlike the Germans, who insist they are keeping their little honeys tightly locked up inside their labs, Japanese molecular biologist Takeo Kubo is certain he'll see his own designer Frankenbees pollinating the cherry blossoms of Tokyo in his lifetime.[13] He turned 59 in 2020.

---

[12] https://www.nationalobserver.com/2020/02/25/news/invasion-frankenbees-danger-building-better-bee (accessed 10 September 2020)
[13] *Ibid*

Rivers Blue and Brown
A River Used to Flow Through It

ON A MAP OF SOUTH Africa, the rivulets, spruits, streams and rivers resemble nothing so much as the capillaries, veins and arteries of a living body. They carry water, of course, water that is the lifeblood of all living systems. There can be no life without water, and there can be no decent living without good water. Earth is the only planet we know of that has good flowing water. But not all of it is so good anymore.

The country's biotic persona is not well. It is a drug addict and its drug of choice is pollution: plastic, fertilisers, chemicals, sewerage – all the waste we generate in our lives, in our factories, on our farms, all the waste we generate as a society and have not got around to dealing with in a mature way. Chuck it in the nearest river, and it is carried away. Who cares where?

There was a time when we could swim in just about any spruit, dam or river, and we did, be it the Vaal or the Crocodile, the Umgeni or the Duzi, the Nahoon, the Great Fish, the Black or the Liesbeek. Now you wouldn't want to put your toe in most of them; they are often afloat with garbage, building rubble, raw sewerage and attendant pathogens, including cholera.

Freshwater ecologists I know reckon maybe only 15 rivers in all of South Africa could be considered as being healthy. How the hell did we, and they – our water custodians – allow things to get so bad?

In a middle-sized town in the Eastern Cape interior, disquiet bubbled up about unacceptable levels of *E coli* in the Great Fish River catchment. The Inxuba Yethemba (Cradock) municipality was at the centre of that bacterial stink. By mid-2016 news of the suppurating discharge had worked its way up the circulatory network to the desk of a responsible person in the Department of Water and Sanitation (DWS).

In October that year the Eastern Cape regional office of DWS issued the appropriate notice to the municipality, demanding it get its wastewater treatment works in order, clean up its act and rehabilitate the affected areas. In response the municipality submitted an action plan to the regional DWS office – where it was rejected on the grounds of not addressing the issues outlined in the original directive.[1]

The DWS issued two more notices to the municipality without any response forthcoming. In April 2018, the DWS dispatched an environmental officer to check out the sewerage works at Cradock. What he found there was disturbing.

The place was locked and abandoned. While wastewater continued to flow into the facility, most of the pumps and aerators were dysfunctional and appeared to have been so for some time. Untreated sewerage was seen oozing from broken sewer lines and manholes. It had only one place to go – downhill into the Great Fish River. Add to this the fact that the town was at the time subject to frequent electricity cuts due to non-payment of its municipal electrical bill. (Although sewerage is water-driven, sewerage processing requires electrical separators and pumps.)[2]

And then something quite unexpected occurred: the DWS instituted a court action against the Inxuba Yethemba authority for non-compliance. The acting town manager replied to a local journalist that one of the problems was that the aerators were putting a strain on the bearings and gearboxes and some other exonerations.[3] (Well, fix it, dear Liza!)

The town with the most widely publicised water troubles in recent times (apart from the 2017/8 near water-outage in Cape Town) is Makhanda, formerly known as Grahamstown. In 2018 it became

---

[1] https://www.engineeringnews.co.za/article/water-department-going-to-court-to-stop-ec-municipality-polluting-great-fish-river-2019-06-25/rep_id:4136 (accessed 20 September 2020)

[2] *Ibid*

[3] https://www.engineeringnews.co.za/article/water-department-going-to-court-to-stop-ec-municipality-polluting-great-fish-river-2019-06-25/rep_id:4136 (accessed 20 September 2020)

one of the first towns – but far from the last – in South Africa to actually run dry.

It was reported that poor people in the City of Saints (courtesy of its many churches) had to buy filtered water at R20 for 5 litres or R5 if they brought their own containers.[4] For many years, authorities had known that the town had major infrastructural problems.

In Centurion, Gauteng, residents established Armour – Action for the Responsible Management of our Rivers – to clean up the stinking Hennops River, once a favourite picnic and swimming spot. They strung a net across the river and, over one weekend in early 2020, 170 bags (about 4 tonnes) of garbage were collected and removed from the river: but it's a never-ending stream of garbage from a settlement upstream. Armour's biggest fear was that a metal weir, which serves to trap most of the solid matter, would be stolen. In May of 2020, the group's Facebook page showed mounds of bags representing 10 tonnes of garbage that had been recently collected.[5]

In 2019, residents of Vereeniging (Emfuleni Local Municipality) took to Facebook to vent their spleen. Pictures posted on the "Save the Vaal Environment" page showed the river horribly polluted. Comments suggested the entire sewage-stormwater system there had collapsed. Drains, pipes and processing plants were clogged with congealed masses of human waste, bacterial mats, all manner of plastic and polystyrene, condoms, nappies and sanitary towels.[6]

It was not the first time the Vaal had been the focus of environmental concerns. Taking its major feed from the Vaal (with some input from the Lesotho Highlands Water Project), the Rand Water Board supplies water to all of Gauteng, which is, as we all know, the powerhouse of our economy.

This is a catchment already highly polluted by coal power stations, gold and coal mining, fuel refining, iron smelting and chemical run-off from intensive farming. (A mining engineer informed me that

---

[4] https://www.newframe.com/makhanda-sues-government-over-sewage-problem/ (accessed 20 September 2020)
[5] https://www.dailymaverick.co.za/article/2019-10-08-citizen-campaigns-to-clean-up-hennops-river-gain-momentum/ (accessed 20 September 2020)
[6] http://www.save.org.za/ and https://www.thesouthafrican.com/news/vaal-river-pollution-sewage-prosecutions/ (accessed 20 September 2020)

most of the vegetables grown in the Vaal catchment contain highly poisonous heavy metals, including uranium.)

ON MY TRAVELS TO RESEARCH this chapter I saw hovels, some nothing more than a hole in the ground covered with plastic sheeting, along the Braamfontein Spruit in Sandton, in open view of golfers swinging clubs at the River Club, one of the most elite golf courses in South Africa. It would be hard to find a more divided society anywhere on this planet.

Along with human settlements spilling over riverbanks and virtually into the rivers when they flood, comes disease. As far back as 2008, cholera outbreaks were reported in Limpopo, Mpumalanga, KwaZulu-Natal and Gauteng.[7] What links just about every river that runs through an urban or peri-urban area is that they are, or were, green belts along which the incoming flood of humanity has found free and open land on which to build (or dig) homes, take water, wash and dump their refuse.

After successive years of drought, in November 2019 the Joe Gqabi District Municipality issued warnings to the residents of Aliwal North, Mount Fletcher, Lady Grey, Ugie, Barkley East, Burgersdorp and Maclear not to drink the water supplied by the municipality without first boiling it. Ten of the 13 towns in the district were restricted to the supply of municipal water for only a few hours a day, in only some cases on consecutive days.[8] Fights over puddles of water and at water tankers became part of the daily grind in towns from Graaff-Reinet to Ugie.

In late 2019 in sultry sub-tropical Durban (eThekwini Metropole), its major river, the Umgeni, made the "polluted" headlines. Users of the lower Umgeni, a popular recreational area, were up in paddles about the continual flow of solid pollution and sewerage spills into the estuary formerly known as Blue Lagoon.[9]

---

[7] https://www.who.int/csr/don/2003_05_23a/en/ (accessed 20 September 2020)
[8] https://www.dailymaverick.co.za/article/2019-11-04-more-towns-running-out-of-drinking-water/ (accessed 20 September 2020)
[9] https://www.iol.co.za/dailynews/news/kwazulu-natal/watch-kayakers-up-in-arms-as-raw-sewerage-flows-into-umgeni-river-30971156 and https://www.

Increasingly, urban pollution and sewerage spills have become the focus of environmental concerns. But chemical and other industrial pollution has long been fouling up our waterways, either by way of stormwater drains or mainlining directly into the fresh channels. Historically, paper mills in places such as Ngodwana, Stanger and Richards Bay have been among the biggest water polluters in the country.[10]

Substantial quantities of chlorine are used for turning wood pulp into paper, with paper mills powered by burning timber or sugarcane "biomass", or waste. Ngodwana, in the confined Elands River Valley of Mpumalanga, was for many years a rampant polluter of the river and the valley's air until being forced to clean up its game.

In September 2019, a spill of industrial oils into Baynespruit, a tributary of the uMsunduzi (or Duzi), had the burghers of Pietermaritzburg in a froth. It was already officially listed as one of the six most polluted rivers in South Africa, when some 1.6 million litres of "mainly edible" vegetable oils and caustic soda had leaked from the Willowton Oil Company (makers of Sunfoil, and Sunshine D and d'lite margarines) and turned the waterway into a toxic grease flow. It also transformed the Duzi downstream (which flows into the Inanda Dam, Durban's main source of drinking water) into a soapy mess.[11]

The spill killed just about all the fish in the river system and a die-off of up to 80 per cent of all river life.[12] Even though the spill had been "really hard", according to a member of the Duzi-Umgeni Conservation Trust, it was only part of the ongoing daily pollution and river degradation, including illegal sand mining in the lower reaches where the Duzi joins the Umgeni.

Positive things to come from the spill, other than focusing public attention on the river's plight, have been the creation of a river disaster fund and discussions between Willowton and local conservation

---

iol.co.za/mercury/news/call-to-curb-pollution-in-umgeni-river-19878736 (accessed 20 September 2020)

[10] https://www.worldwildlife.org/industries/pulp-and-paper (accessed 20 September 2020)

[11] https://www.dailymaverick.co.za/article/2019-08-27-rush-to-save-kzns-umsunduzi-river-after-disastrous-toxic-spill/ (accessed 20 September 2020)

[12] *Ibid*

groups aimed at establishing a Baynespruit conservancy.

Back in the 1950s, urban planners in Cape Town came up with a bright plan to deal with their rivers that would run dry in summer and then flood in winter.

First, they deepened and straightened the river channels, and then they cast them in concrete straightjackets. That done, they could crowd people onto the otherwise bleak Cape Flats and be rid of those troublesome rivers: the Liesbeek, the Black, Elsieskraal and the Salt which flow into Table Bay; also the Diep, the Keyser, the Sand, the Lotus and the Langvlei, which flow into the surprisingly extensive lake system feeding into False Bay.

And now these rivers are contaminated with sewerage. When quizzed about restoring the concrete Sand River channel, someone with a conservation portfolio in the area asked me: "Why, when they work so well to get rid of stormwater?" Suffer the children of those areas, who have nowhere else to play.

"STATE CAPTURE" IS NOTHING NEW here or indeed in many – if not most – countries in the world. In colonial times, the Empire owned you pretty much wagon, horse and hay. In South Africa, books such as *The Super-Afrikaners* opened our eyes to just how thoroughly the Afrikaner nationalist movement had taken control of every aspect of the country's functioning.[13] The ANC did not capture our railway system: The Nats did it a century ago. But one thing those hoary old white men ensured was that, at least as it impacted on the white population, the country's infrastructure ran as well as did its rail network.

But, starting sometime after we were gifted democratic government, the stories – not unfounded, it turned out – about how the country's infrastructure was breaking down (potholes being the topic most favoured in this discourse) began. In fact, the whole picture was much larger, and more complex. You can say it started with state corruption and nepotism that, over time, we witnessed ramping up to eye-watering levels never seen before.

---

[13] *The Super Afrikaners* by Ivor Wilkins and Hans Strydom (Jonathan Ball, Cape Town, 1978, 2012).

The first I became aware of it was when my city, Cape Town, began to have its assets stripped, starting soon after Nelson Mandela retired from official office. Then came our "arms gate", but even that turned out to be just small change. During the presidency of Jacob Zuma (enter the Gupta cabal) we stood by and watched, our figurative jaws drooping to the ground, as our country was economically ransacked.

Virtually every government department – including defence, education, police, communications, railways, you name it – was involved in the plunder. Also all the state owned entities, or SOEs, including our national air carrier SAA, broadcaster SABC, and power utility Eskom. And among them is the Department of Water Affairs and Sanitation. Under two successive ministers, that department was stripped down to its skeleton and actually declared bankrupt in 2017. (It was R4.3 billion in debt.)[14]

Things really did start to fall apart. When the Zumarite faction eventually got its hands on our internal revenue service, it looked like it might be game over for our economy and, by extension, our everything. (Luckily, it looks, to me at any rate, that this story might yet have a silver-lined ending.) But where did all that money go?

There is mounting evidence (in news and official reports, books, accusations and counter accusations, charges and commissions about government corruption, right up to the Zondo Commission hearings) that much of the money "stolen" over the past five years went to funding election campaigns.

A watershed moment took place in 2017 to decide who would succeed Jacob Zuma as president, and whether the "capture" would continue or not: the one preferred by the president and his followers, his ex-wife Nkosazana Dlamini-Zuma; Zuma's political opposite and vice-president Cyril Ramaphosa; or wild card Lindiwe Sisulu (water minister at the time). Once that was settled came the battle over who would sit on the vice-president's throne. It was difficult to calculate, as there were many different accounts of money being used to buy influence and votes.

---

[14] https://ewn.co.za/2017/02/12/report-water-and-sanitation-department-bankrupt (accessed 20 September 2020)

If *Times Live* and the *Mail&Guardian* are to be believed, around R500 million of it came from water and sanitation to fund the political aspirations of Sisulu. Denials and accusations of "fake news" (sound familiar?) poured like floodwaters from that office. But the reality remains: The people in many places – including Makhanda, Graaff-Reinet, Beaufort West and Giyani – have virtually no access to clean drinking water. (As I write this, Port Elizabeth stares into the void of water "day zero".) Many of our rivers are like open sewers; vital new dams and other water infrastructure stand incomplete. There's literally no money left to fix any of it, or even do the most basic maintenance. It's a dirty game, and it's called politics.[15]

But, while all this was going on, there were other, even stronger, forces at play: urbanisation of unprecedented magnitude, driven by the twin turbocharger of political freedom and economic hardship. It's a formula for social, economic and biological meltdown.

Another place to start this story of "breakdown" is with the old Biblical observation that as you sow, so shall you reap. The sowing for 100 years or more of monocultural, unpigmented privilege is reaping an untidy, multi-coloured harvest. Gravity dictates that things flow downhill, into gaps and traps and receptacles and, geographically speaking, dongas and valleys and rivulets and spruits and streams and rivers. Put another way: If you have even a scant knowledge of physics, you'll know the second law of thermodynamics tells us that, in order to keep a system intact, increasing amounts of effort (energy, heat, money, work) must be invested. If you don't, things fall apart. It's as simple as that: it's called entropy, and it is a central force governing how the universe works.

In the years prior to 1994, when only 30% of the population seemed to matter to Government, much energy, heat, money and work was continually invested in the country's various systems, and they were running pretty well. The remaining 70 per cent were living in "township" ghettos with few if any tarred roads, potholed or not.

---

[15] https://www.timeslive.co.za/sunday-times/news/2019-11-17-lindiwe-sisulu-accused-of-using-budget-to-run-for-anc-deputy-president/ and https://mg.co.za/article/2019-11-15-00-complaint-sisulu-in-bid-for-presidency/ (accessed 20 September 2020)

Most did not have running water, electricity or modern sanitation, and much else that was enjoyed as a moral entitlement by the 30 per cent – including decent education.

Then, overnight, all those things became enshrined as basic human rights, and suddenly a system built to deliver services to around 15 million people had to be retooled to serve some 50 million, more than tripling the country's infrastructural load. Little wonder it has come under strain.

"The reality [of the apartheid legacy] is that so many of our white compatriots simply want to deny," writes political analyst Oscar van Heerden in the *Daily Maverick*. "They would very much like to believe the narrative that blacks inherited a strong economy and a stable country with excellent infrastructure that worked."[16]

Our reality in the early 1990s could not have been further from that truth, he argues. "Municipalities worked because they only worked for a small preferred minority of white citizens. Electricity, water and sanitation were mainly for whites only. So, whites conclude that blacks cannot govern, they were corrupt and look at how *they* have destroyed this once beautiful country of ours."

UNTIL FAIRLY MODERN TIMES CITIES and large towns the world over were wallowing in human and animal waste. The combination of manure from horse-drawn carts and carriages and human slops thrown out of windows into the streets below made perambulating a hazardous undertaking.

The pavement (the paved side, or crossing stones that were raised above the road) was a Roman invention, in order to keep the shoes and sandals of pedestrians dry and clean. It was only the invention of the internal combustion engine in the late 19th century that solved the problem of cities literally drowning in horse apples (as they were called in delicate society).

The summer of 1858 was a particularly hot and dry one in London, the largest and most powerful city in the world at the time, engine house of the British Empire. Starting in July, a "great stink"

---

[16] www.dailymaverick.co.za/opinionista/2019-11-13-south-africa-25-years-on-slowly-slowly-catchee-apartheid-monkey/ (accessed 10 September 2020)

enveloped the city. The Thames was running lower than normal, and a combination of human waste and industrial effluent had accumulated on the exposed mud banks where it festered. Normally, even though all the city's waste emptied directly into the river, its strong flow would clear the mess out into the English Channel.

But an antiquated and inadequate sewerage system had burst its seams and, combined with the heat, resulted in central London being enveloped in the infamous malodourous pall. Residents, including much of the medical fraternity of the day, believed it was a "miasma" rising from the effluent rather than, as we know now, the filthy infected water that was the cause of an outbreak of cholera.

In reaction to the Great Stink, one of the engineering marvels of the age, Crossness Pumping Station, was built. Designed by architect Charles Henry Driver and engineer Sir Joseph Bazalgette, it has been described as "a masterpiece of engineering – a Victorian cathedral of stone and ironwork".[17] It was located at the eastern extreme of the city, close to the river estuary.

Four huge steam-driven pumps (built by James Watt & Co and named Victoria, Prince Consort, Edward Albert and Alexandra) raised the incoming effluent up about 10 metres and then discharged it on the outgoing tidal bore. The population of the city continued to swell, so in 1897 four extra pumps were added to cope with the incoming flood of human waste. In 1913, the steam engines were replaced with diesel ones. In 1953 Crossness sewerage works was mothballed and replaced by numerous more modern treatment facilities.

The detail of the Crossness station, now a restored and listed building, is given here merely to illustrate that whatever your current travails and concerns, you are never alone in time or place. We just hope that our political and sanitation systems can be fixed before another "great stink" – replete with cholera outbreaks – envelops us all.

But enough with moaning and groaning – how can our water-sewerage problem be fixed? The conventional knee-jerk response is

---

[17] The citation is widely attributed to Sir Nikolaus Pevsner, a German-British art historian. See for instance, https://timmyatt.com/tag/nikolaus-pevsner/ (accessed 28 August 2020)

to send in the engineers with heavy machinery and bulging budgets. But there is another way, call it the green way. The starting point for this is the new green-deal trifecta of re-use, recycle and reduce.

You cannot easily reduce the amount of human waste you produce daily, but the amount of water you use to dispose of it (actually passing your problem on downstream) you certainly can. About 45 per cent of potable municipal water consumed by the average urban household is used for flushing toilets.

Human waste is a very dirty and smelly business, so how else do you deal with it? One way is a system referred to as off-grid sanitation. That's how it's widely been done in rural areas, usually with long-drop toilets, forever. In urban areas, various low- or no-water toilets have been and are being tested.

Around the time this book was written, several bio and tech organisations in South Africa, with support from the Bill and Melinda Gates Foundation, launched the South African Sanitation Demonstration Programme to transform the way we treat human waste.[18] Very simply, you give it value: If someone can make money out of it, it will sell. Some of the products that can be produced from human waste include bio-gas, bio-char (charcoal), oils, fertiliser and proteins.

Waterless toilets are nothing new, but getting them into our homes in a way that is aesthetically acceptable is still some way off. Nevertheless, along with the green, there is a brown revolution coming our way soon. Because if we don't go down the brave new brown road, we'll all soon be up to our necks in yellow bricks of sewerage. As they used to say in ancient Rome: *sumus semper in excretum, sed alta variat*.[19]

---

[18] www.susana.org/en/knowledge-hub/projects/database/details/239
[19] We are always in the poo, it is only the depth that varies.

Plastic O Plastic
Macro, Micro and Now Nano – It's Time to Clean Up Our Act

We all know that plastic has become one of the leading causes of environmental concern, aesthetically on land but, more crucially, biologically in our waters.

Given the fact that many readers will already be suffering advanced PCFS (planetary-collapse fatigue syndrome), following are some fun facts about plastic to get you into the spirit of it:

- Nearly half of all the plastic produced in the world today is used just once then thrown away.
- Every year around 9 million tonnes[1] of plastic ends up in our oceans.
- That's enough to place about five plastic shopping bags filled with plastic trash on every 30 centimetre stretch of coastline right around the world (coastlines being extremely problematic to measure).
- Less than one-fifth of all plastic produced is recycled.
- A trillion (1,000,000,000,000) plastic shopping bags are produced worldwide every year, each with an average useful life of 15 minutes. The material will persist for several centuries or longer.
- Plastic (from plant cellulose) was invented in the mid-19th century.

---

[1] For the record, a tonne is the metric measure for 1,000 kilograms. A ton is the Imperial measure equal to 2,240 pounds (1.016 tonnes); it comes from tun, the largest size wooden cask used for shipping liquids in days of yore.

- Educated estimates say plastic kills millions of marine animals every year.
- Every minute around the world, one million plastic bottles are purchased. Almost all of them will be thrown away. That's 52,6 billion a year.[2]
- Around 25 billion shoes containing plastic-rubber composites are bought each year, mostly slip-slops and cheap trainers, most of which end up as litter.
- Up to 6 billion plastic toothbrushes are thrown away annually.
- An estimated 400 million car tyres are littered around the world.
- About 3 trillion cigarette butts are tossed every year. Those with filters, the vast majority, pose a toxic risk to aquatic animals when they end up in rivers and seas looking like food.
- The typical woman who reaches menopause will have discarded around 10,000 menstrual pads or tampons, plus the applicators, along with the associated packaging, in her lifetime.
- Since mass production began around 1950, we have accumulated around 9 billion tonnes of plastic, most of which will still be around long after we are not.
- Most of the plastic that washes up on our beaches, as opposed to that which is dropped or comes via stormwater drains and rivers, originates in Indonesia.

IF YOU WERE LOOKING FOR a supermodel to showcase the horrors of plastic pollution, you'd need search no further than the olive ridley sea turtle with a plastic drinking straw lodged up one of its nostrils. The eight-minute YouTube video clip has had more than 23 million views.[3] A marine biologist, using the pliers of a multi-tool, finally manages to extricate it from the crying turtle's bleeding nose.

---

[2] These figures, millions and billions, can be confounding. However, to get some kind of handle on them, consider this: to count a million seconds (or rands, or bottles) would take about 12 days if you counted at the rate of one unit a second; to count a billion would take more than 30 years.

[3] https://www.youtube.com/watch?v=lVPSTkYihCY (accessed 29 August 2020)

The entire procedure was filmed off Costa Rica by marine biologist Christine Figgener. Be warned, it is highly distressing to watch to the end. Admittedly, one turtle does not a universal meltdown make, but like the cuckoos and doves and nightingales that have all but disappeared from English country gardens, the global disappearance of marine creatures is now our canaries in the plastic mine.

Seabirds, including almost all albatrosses, which snatch up fish and squid from the ocean surface, are most at risk. On nesting islands off Australia and New Zealand it has been discovered the flesh-footed shearwaters, which nest or roost there, have the highest percentage of plastic to body weight of any marine animal. An astonishing 90 per cent of fledglings were found to have ingested some kind of plastic.[4]

In some cases, with adult birds particularly, that would not be an automatic catastrophe, but with young birds it is, for two reasons: first is that they think it is food and so suffer malnutrition with stomachs full; the other is the currently unknown effects of all the chemicals in, and added to, the plastic confetti they are ingesting.

Way back in 1856, a man named Alexander Parkes in Birmingham (the original one, in England) patented a product he called Parkesine. The plastic material (from the Greek *plastikos*, capable of being moulded or shaped), made from plant cellulose, was shown in London at the Great Exhibition of 1862, where it won him a bronze medal.

For most of the rest of that century, it was used as a replacement for elephant ivory in piano keys and billiard balls, bringing those now cheaper leisure pursuits into ordinary people's homes. But, other than that, the new substance did not really catch on until the next century.

Early in the 20th century, a new mouldable material called Bakelite was synthesised from phenol and formaldehyde. It very quickly became the new wonder material. It was used for handles, knobs, pipes, rods, in sheet form and covers for many things including telephones and radios.

In the first half of that same century, chemists working for large petroleum companies discovered that waste gases going up in smoke from refinery stacks, stuff like ethylene, could be used as the building

---

[4] https://www.stuff.co.nz/environment/94448787/third-of-seabirds-found-dead-on-nz-and-australian-shores-had-eaten-plastic (accessed 20 September 2020)

blocks (monomers) for a wide range of plastics (polymers) such as polyethylene terephthalate, or PET.

These new "magic" products ushered in a range of uses not before contemplated. They quickly replaced wood, metal, stone, bone, glass, leather and ceramics in many applications, including our cellphones, laptops and television sets. It has been said that plastic was one of the weapons that helped the Allies win World War II, as it was used as a substitute for otherwise scarce resources in everything from guns to aeroplanes.

One thing is for sure, no matter how much some people would like us to be rid of them, plastics are omnipresent and far too useful to be going away soon – other than into our rivers and oceans, that is. The American inventor John Wesley Hyatt (1837–1920), who simplified the industrial production of plastics, believed they would eliminate the need to "ransack the world of substances which are growing constantly scarcer".[5]

He was probably right about that but did not envisage the knock-on. Take your home, for example: If you are an average person (and there is every reason to believe you are), at current figures you are adding about 50 kilograms a year to the world's load.

Starting in the bathroom, there's your toothbrush (unless, like mine, it's bamboo with natural fibres), the toothpaste tube, towels if they are polycotton, cosmetic tubes and pots, shampoo and conditioner bottles, various containers, tampon packaging and applicators. There's even more out of sight, like most of the plumbing, and probably much of the fittings, including your toilet bowl.

In the bedroom there's the bedding (unless you insist on 100 per cent cotton, or silk), the mattress and pillow foam, the carpets, most of the shoes in your closet, and a large proportion of your clothing. Possibly the cornice detailing in the room. Even the active, green, veggie-eating, turtle-hugging readers among us – just about every

---

[5] In an article in *Smithsonian Magazine* posted 1 July 2010, Jesse Rhodes writes that this text appeared on a promotional pamphlet for the celluloid Hyatt and his brother started producing in the late 1800s. See https://www.smithsonianmag.com/smithsonian-institution/american-history-highlights-celluloid-and-the-dawn-of-the-plastic-age-139344351/ (accessed 29 August 2020)

item of sports clothing you own is made with largely synthetic fibres.

In the kitchen there are jugs and all those plastic containers that don't have lids, and the lids that no longer have containers, spatulas and other cooking implements, the non-stick layers on your pans, food wrapping, spice bottles, the vacuum cleaner covering and pipes, garbage bags, even bags we use for recycling, how ironic.

In the garage it's not just parts of your car (around 20 per cent) and its tyres, but parts of your bike, golf clubs, tennis racquets… There is virtually no aspect of modern life, home, work or industry where plastic does not prevail: The list is way too long to expand. Just how do we deal with it is the question though – and the problem.

Sometimes the microcosmic scale serves best to illustrate the macro view, in this case of worldwide plastic pollution. In 2018, when Greenpeace organised a clean-up of Freedom Island (a critical habitat for migratory birds) in the Philippines, they found six international brands were responsible for more than half the total of their trash haul.[6] Just about all of us know them well. We buy their chocolates and ice creams, soaps, cosmetics, pet foods, biscuits, bottled water, baby foods, teas and coffees, cereals and much more.

If you want to watch some hard-core garbage porn, Google the YouTube videos showing "trucks tip garbage into river in India". It's like watching Mother Earth being violated. In that country, India, more than 40 per cent of all plastic is used for packaging that will be tossed as soon as the contents are opened.[7] Hello, rivers!

A map of where all the plastic pollution in our oceans comes from reveals that 10 river-mouth hot spots – most in Asia, two in Africa and one in South America – are responsible for more than half of it. They are (in no specific order) the Amazon (Brazil), Niger (Nigeria, as well as Port Harcourt, the oil-producing centre of Nigeria), Pasig (Philippines), Brantas (Indonesia), Mekong (Vietnam), Irrawaddy (Bangladesh), Ganges (India), Pearl (China, Hong Kong), Yangtze

---

[6] https://www.greenpeace.org/usa/compass/spring-2019/gp0str3eo/ (accessed 20 September 2020)
[7] https://www.nationalgeographic.com/magazine/2018/06/plastic-planet-waste-pollution-trash-crisis/ (accessed 20 September 2020)

(China) and the city of Shanghai.[8]

Just five Asian countries produce more than half of all the plastic waste going into our oceans: China, India, Indonesia, Vietnam – no surprises there but, surprise! – Sri Lanka.[9] Even if you collected and disposed safely of all the plastics produced and used in Europe, the Americas, Australasia and Africa, it would make only a dent in this pile.

A staggering 360 million mostly non-returnable plastic bottles are sold globally every day. That's around 1.3 billion a year.[10] But at least they are, theoretically and potentially, recyclable or re-usable. Not so sachets. Those little plastic "pikkies" are used to vend single portions of instant coffee, shampoo, toothpaste, spreads, sauces and also the energy gels so fancied by runners and cyclists. Because they are not recyclable, they have no monetary value, which makes them especially troublesome for models that seek to contain the plastic plague through recycling and repurposing.

But that's not even the worst of it. For years, the people who measure these things were perplexed by the fact that, no matter how much they collected and counted, they could not make the scales balance between what they knew was being produced and what was ending up in tips, rivers, along beaches and in oceans. Where was the shortfall?

The answer came in a landmark paper published in 2004 by Plymouth University professor of marine biology, Richard Thomson, which was a high-water mark in a messy field of study. Thomson spent several halcyon years on the Isle of Man, where he had teams of students look for all the small pieces of plastic they could find. While plastic mostly does not go away, especially in sunlight, it becomes brittle and breaks up into increasingly small particles. Thomson's students found a great deal of it, much of it

---

[8] "Plastic" in the June 2018 edition of *National Geographic*, pp 40–69. Also see https://i.pinimg.com/originals/3a/f3/ec/3af3ec5f4f1593718074dcb33689836b.jpg (accessed 20 September 2020)

[9] *Ibid*

[10] https://www.nationalgeographic.com/environment/2019/08/plastic-bottles/ (accessed 20 September 2020). To get a handle on what these numbers mean, if you tapped a drum once a second, it would take you about 12 days to tap a million times. To tap a billion times would take more than 30 years.

only the size of the sand grains, on the beaches.[11]

By crunching the numbers, Thomson found the missing plastic and in the process coined the term that we have all come to know well and dread – microplastics. Find a beach anywhere in the world and you'll find them there, as well as on the highest mountains, on the deepest ocean floors and even floating in Arctic ice.

The largest mat of floating plastic trash lies between Hawaii and the North American coast. It covers an area about three times that of the British Isles. Most of it does not originate from either region but is gifted to the Aloha state and surrounding ocean by the North Pacific gyre, which gathers up pelagic trash from all around southeast Asia and the associated islands.[12]

It is estimated that as much as 15 per cent of the lovely "sand" of the legendary beaches of Big Island in Hawaii is actually microplastic. While on assignment there, *National Geographic* staff writer Laura Parker recalls, they "crunched like Rice Krispies under my feet".[13]

Ocean currents are mysterious and circular; everything is connected. In 2015, two Australian scientists cooked up a scheme to see if they could find plastic – and if so how much – on one of the remotest, most uninhabited places on Earth.

They chose tiny Henderson Island, in the Pitcairn group in the southern Pacific Ocean, and spent three months there collecting and counting plastic. You might recall that was where the *Bounty* mutineers chose to hide from the omnipresent and vengeful Royal Navy. Anyway, the two Aussies calculated the island had around 18 tonnes of visible plastic waste (they did not measure the microplastics), the highest count ever recorded by area anywhere.[14]

As if microplastics were not bad enough, now we have to get our heads around nanoplastics – the things that exist somewhere above

---

[11] https://www.plymouth.ac.uk/alumni-friends/invenite/issue-2/the-big-interview-professor-richard-thompson (accessed 20 September 2020). In 2018, Thomson was awarded an OBE for his services to marine science.

[12] https://www.nationalgeographic.com/magazine/2018/06/plastic-planet-waste-pollution-trash-crisis/ (accessed 29 August 2020)

[13] *Ibid*

[14] https://www.dailymail.co.uk/sciencetech/article-7300649/Plastic-junk-spawns-desert-island-Pacific.html (accessed 20 September 2020)

molecule level but mostly out of sight, unless you happen to be a water flea or oyster.

The base of the marine food system is plankton, tiny particles of plant or animal matter. When plastic breaks down to really tiny particles, it resembles plankton, and just about all particle-feeders gobble them up. They are like the grains of sands on the beaches or stars in the sky.

A microscopic image of a transparent water flea just 3 millimetres long, published in the *National Geographic* article already referred to, shows a noodle soup of tiny nanoplastic pieces. A study of 114 fish markets around the world found similar filaments, fibres and microbeads in just about every species, from Olly Oyster and Shrimpy Shrimp right up the food chain to Titanic Tuna.

An ultra-sensitive sound recording detected one of them saying, "thank you humans for all your exfoliants..." Not really, but they might as well have. Microbeads from cosmetics form a large portion of micro- and nanoplastic mass. No one yet knows what effects they will have in the long term, but there's no reason to think it will be good.

There are the pessimists who think we are doomed to destroy nature on our planet, and there are the optimists who think our clever technologies, those that got us into this pickle, can be used to get us out of it. The jury is still locked in debate.

An Irish schoolboy named Fionn Ferreira won first prize at the 2019 Google Science Fair for his method of removing micro- and possibly even nanoplastics from the sea. The smart 18-year-old figured, if like attracts like as he was taught, then some kind of plastic solution could be used like a magnet to attract the tiny plastic particles.

What he came up with was a liquid called ferrofluid, made from petroleum products, that does just that. "I'm not saying that my project is the solution," he told *The Irish Times*: His ferrofluid magnet recovered 88 per cent from a water sample. "The solution is that we stop using [non-recyclable] plastic altogether."[15]

Every year an estimated 11 million tyres are sold in South Africa,

---

[15] https://www.ecowatch.com/google-science-award-irish-teenager-2639623184.html?rebelltitem=1#rebelltitem1 (accessed 20 September 2020)

a small portion of the 1.6 billion worldwide admittedly, but they are ours. Most of us have a vague idea where our old tyres end up (littering the countryside mostly, also on township cars as "new re-treads"), but have you ever thought about where all the tread from those old tyres goes? To the road verge, down stormwater drains, into rivers and finally into the oceans, broken down into micro- and nanoplastics.

A partial solution is being piloted in various places around the world, including South Africa, using plastic waste to surface roads. The first test trip was laid in Jeffrey's Bay in early 2019.[16] Not only could they use up otherwise unusable plastic waste, but plastic waste also makes for a less abrasive surface and therefore causes less slewing of tyre tread. That's the theory anyway, but the tyre companies won't like it. Remember, Thomas Edison made a light bulb that could last a century, but that did not make for good business.[17]

Another formative plastic repurposing project is called Eco-Blocks (as opposed to eco bricks, which are plastic cold-drink bottles stuffed with soft throw-away plastic waste). In the case of the blocks, waste plastic is reduced to something called pre-conditioned resin aggregate, which is then moulded into building blocks stronger than conventional cement ones. Their commercial production in South Africa is still some way down the line, but we do have numerous companies already making timber-replacement products from waste plastic. Let's hope they catch on.

South Africa was once ahead of the litter curve. In 2003, then Minister of Environmental Affairs Mohammed Valli Moosa banned single-use plastic bags – at least he tried to. Unfortunately, the push-back from the bag makers was stronger, so that did not work out so well. But it is the reason we have to pay some small change for them at supermarkets. But not, alas, the thinner barrier bags into which the till packers in many supermarkets insist on putting every other item,

---

[16] https://businesstech.co.za/news/motoring/333307/work-begins-on-south-africas-first-plastic-road/ (accessed 20 September 2020)
[17] en.wikipedia.org/wiki/Incandescent_light_bulb (it is a long read, but the pertinent section is under heading "Light Output and Lifetime" – GE was Edison's company that aimed to reduce the lifetime of incandescent bulbs)

before popping them into your other bag.

Here and indeed worldwide there is little incentive for the companies making throwaway plastic products to reduce their wares along with their profits, their argument being that it will lead to job losses. That is typical corporate smoke screening, but a strong financial lobby always seems to be able to persuade a politician to do the wrong thing.

People who would lose the most are those at the very top of the pile. Women like Neli Xulu. She is one of thousands who scour tip sites in towns across the country for waste they can recover and sell to recycling companies.

Xulu lives in a shack made of plastic sheeting and cardboard on a site sandwiched between the New England dump in Pietermaritzburg and the Msunduzi River. There are around 300 people living in Ehlathini, with more shacks mushrooming all the time. The people of Ehlathini (it means forest, or bush) share just one communal water tap and a few toilets, or do their private business in the bush.

In Kranskop, where Xulu comes from, she earned R600 a month as a domestic worker. As a waste picker, she is able to provide for her own two children as well as four of her late sister's by earning R3,000 worth of scrap in a fortnight, working from before dawn until after dusk.[18]

And just then your Uber Eats delivery arrives: "One deep fried calamari with bottle tops, two hake and salads with extra chip packets, and two mussel pots with microbeads. Thank you for ordering from Davy Jones's Locker."

AFTERTHOUGHT: ONE OF THE MANY things we'll remember from COVID-19 is the vast amount of medical and other protective gear that will have been discarded, much of it – like latex gloves – after only one use. Of all the precautions we took with the Corona virus, what we did with the waste it generated was not one of them. Compounding the problem was the fact that during lockdown time even waste recycling plants were closed. Maybe we'll rethink this one next time.

---

[18] https://www.groundup.org.za/article/people-flock-live-dump-site-pietermaritzburg/ (accessed 29 August 2020)

The A, B and C of Climate Change
Anguish for the Atmosphere, Butchery of the Biosphere, Crying for the Cryosphere

THE CRYOSPHERE

FOR STARTERS, FOR CRYING OUT loud, what is the cryosphere? The word derives from the Greek *kryos* – which can mean cold, frost or ice – and, of course, sphere. A sphere can mean different things, like a globe such as the Earth, or an area, as in a sphere of influence. Cryosphere is the term used collectively to describe all the areas of our planet where water is to be found in solid – that is frozen – form. It includes sea ice, lake ice, river ice, snow cover, glaciers, ice caps, ice sheets and frozen ground, or permafrost.

Fine, but what has the cryosphere got to do with us here in South Africa? The thing is, as Alexander von Humboldt taught us, everything is connected. If you are indeed sentient, you'll know glaciers in the Himalayas, the Alps, Alaska and Greenland have been retreating, melting, for some time now.

In Africa, the ice and snows of Kilimanjaro, Mount Kenya and the Rwenzoris are all but gone. The extent and thickness of the sea ice of the Arctic has been dwindling since at least the 1960s and even the massive Antarctic ice shelves are beginning to break apart. It's all happening while you read this, and you can bet your last cent that it will have an impact on you, massively.

The most immediate bump-on of the warming and melting ice is that sea levels will rise, and the homes of countless millions of people living in low-lying coastal areas will begin to get waterlogged. That's still the easy part: People can pack up and move to higher ground, and they will. (The results will not always be easy, or pretty, but move they will have to.)

Even more serious than the glaciers and ice sheets melting, though, is the situation with the world's permafrost regions that cover vast swathes of North America, most of Greenland, northern Scandinavia, the entire Qinghai-Tibetan Plateau and all of Siberia. We are talking about an area about 19 million square kilometres in extent.

Clever climatologists who measure things like polar ice cores, glacial retreat, sea-floor sediments and the like, inform us that the Arctic permafrost has been melting for the past two centuries, starting slowly at first, but the rate is now increasing. Permafrost is frozen soil and, like soil everywhere, it holds carbon in large amounts. As the permafrost melts, it has the potential to release as much as 1.6 trillion tonnes of carbon, in the form of methane gas and carbon dioxide, into the Earth's atmosphere.[1]

That amount of carbon gas dwarves all the emissions currently being released from all the fossil fuel vehicles, aeroplanes, ships and power stations around the world. But don't think they are not connected: It's the human-fuelled greenhouse effect that accelerated the climate change curve in the first place and that is, mostly, causing this permafrost "big warm".

Another thing about good old Mother Nature is that all her processes are concentric; poke her with a stick here and she swings around and bites you there. We pump carbon gases into the atmosphere, the Earth warms up, the glaciers and permafrost begin melting at accelerating rates, which releases more carbon gas, the rate of warming increases…

Another loop in the global warming feedback scenario is being played out in our oceans. They are responsible for absorbing the

---

[1] https://www.dailymail.co.uk/sciencetech/article-8550123/Permafrost-melt-release-40-BILLION-tons-CO2-atmosphere-previously-thought.html (accessed 20 September 2020)

whale's share of all the CO_2 we are spewing out – so far as much as 30 per cent of what our various combined activities have produced.[2] But, not only does this mean their absorption abilities reduce (a sponge can hold only so much), the increased carbon dioxide creates increased acidity, and that is bad for things living in the sea, most especially plankton and coral reefs and anything else with a dissolvable calcium-carbonate shell.

If we don't stop spewing out all this carbon, we're screwed. The end game is not a pretty picture. It's still a wee bit premature to make the call, but we are likely to kill ourselves through simple stupidity. It might be nuclear, it might be viral, but more and more people are putting their bets on the black numbers of environmental implosion.

Most people way down here in Africa would not have heard of an organisation called the International Permafrost Association (IPA). But a surprisingly large number of people live in permafrost regions (places where the mean annual temperature is below freezing). For them it's easy to build on frozen soil, but what to do when the very foundations of your home begin to ooze away?

One fun part of the IPA's preoccupations is seeing what once-living beasties pop out of the thawing ice. We know mammoths have come to light, and that a small group of Russian scientists may or may not even have enjoyed a mammoth steak braai. Some other things found include antediluvian (before the Flood) homes made from the tusks of the woolly pachyderms. Also, plants and seeds, including a 30,000-year-old plant specimen of *Silene stenophylla*, which was found down an ice-age squirrel burrow, and revived.

Anyone who has watched the chilling crime-thriller TV show *Fortitude* will remember how the polar series climaxes with the discovery of the thawed body of a diseased mammoth carcass. It was not just the stuff of TV thrillers: Several years ago, an anthrax outbreak on the Yamal Peninsula in Siberia was thought to have been caused by exactly that.[3]

---

[2] https://www.weforum.org/agenda/2019/03/oceans-do-us-a-huge-service-by-absorbing-nearly-a-third-of-global-co2-emissions-but-at-what-cost/ (accessed 20 September 2020)
[3] https://thebarentsobserver.com/en/arctic/2016/08/scientist-yamal-

It was George Bernard Shaw who observed that, in time, every joke ends up being serious. In the same way all good science fiction eventually becomes reality. Think of all the fun we have to look forward to.

THE BIOSPHERE

IN HIS ECOLOGICAL GEM *The Song of the Dodo*,[4] David Quammen likens the biosphere to a delicately woven Persian carpet (as alluded to previously). During the 20th century we have repeatedly taken shears to the carpet and diced it up into ever-smaller pieces. We have connected cities to towns and back to cities with highways, byways, railways, enormous industrial farms, and villages that spread into neighbouring villages. Where once we had an ever-changing, kaleidoscopic, tangled riot of textures and colours, we now have geometric uniformity where nature huddles in peril. And now it's made of rayon, not wool.

Take the case of the primates – chimpanzees or baboons, for example. If you look at the numbers, both species appear to be safe enough, shielded from the predations threatening most of the other large African mammal species. But mere numbers don't explain what is really happening on the ground.

Chimpanzee females are driven out when they reach sexual maturity, so that they can go and spread their genes among neighbouring troops. With baboons, the males are the ones who must go and sow their seeds far and wide. In the former stronghold of chimpanzees of the Gombe-Mahale hills in Tanzania, the strip of forested hills has been so closely encroached by farmlands and villages, the famous chimps there no longer have natural corridors along which to migrate.

Closer to home our own primates are suffering similar scenarios. Not so long ago the several troops of chacma baboons living on the Cape Peninsula could roam from Table Mountain to Cape Point almost uninterruptedly. However, from the 1990s, things changed

---

anthrax-outbreak-could-just-be-beginning (accessed 20 September 2020)
[4] *The Song of the Dodo: Island Biogeography in an Age of Extinction* by David Quammen (Pimlico/Random House, London, 1996)

dramatically for them.

What have become almost annual wildfires continually lay waste to their natural feeding areas, forcing them to seek sustenance down from the mountain slopes. At the same time, creeping upwards has been a slow and steady rise of the urban footprint. The national park has been effectively cut up into a number of discrete pieces, which greatly inhibits migration.

Where can the baboons go now, when things get dire, but into the gardens of homes of the upwardly mobile? Not only baboons, also caracals, genets, porcupines, tortoises and snakes, and all the other animals have to run the gauntlet of human persecution. (The organisation Baboon Matters attempts to monitor and diffuse human-baboon conflicts, so that other organisations do not step in and cull the "rogue" primates they are trying to save.)[5]

In sultry Durban, an urban area that was long ago hemmed in by sugarcane fields, hacked out from the indigenous coastal bush and forest. Here, as in the Cape, the few decades around the turn of the 20th century have seen the old city spill out into the cane fields and expand mostly northwards into KwaMashu, along Umhlanga Ridge, through La Lucia to Umdloti and Shaka's Rock, amalgamating Ballito, Umhlali, Salt Rock and advancing towards Sheffield Beach and Blythedale.

A generation ago those places were isolated holiday villages with ample open space and wooded areas. Now that it is one almost solid built-up area, where can the vervet monkeys go to feed? Into gardens and pantries, raising the ire of the newcomers – those pesky humans. It's the same story the world over.

In South Africa, SANBI reckons one in seven of all our indigenous plants and animal species is in mortal danger – those we have not already lost.[6] While about 60 per cent of our biota is well protected in national parks and nature reserves, 15 per cent is threatened with

---

[5] baboonmatters.org.za
[6] https://www.dailymaverick.co.za/article/2019-10-08-one-in-seven-plants-and-animals-in-sa-threatened-with-extinction-new-report-shows/ (accessed 20 September 2020)

extinction.[7] Wetlands and river estuaries are particularly vulnerable ecosystems, yet the least protected of all our natural habitats.

The 700-plus alien plant species that have gone feral pose another major threat to the natural biota. These include plants introduced to beautify our gardens, such as lantana, bugweed, Madagascar periwinkle, morning glory and Jerusalem thorn among many. Numerous species of pines, alien acacias and gums (eucalyptus), originally introduced to provide timber, now pose the greatest threats, as do smaller Australian acacias that were introduced to help stabilise drifting coastal sands, namely Port Jackson willow and rooikrans.[8]

The threats are many, including silver and black wattles that invade river banks, smothering indigenous riverine plants and destabilising the banks; stands of invasive (as well as plantation) trees that help fuel and greatly intensify wildfires; plants including wattle and prickly pear that invade grazing lands; and all the trees that suck up precious water in an essentially arid country and, by outcompeting ground-cover plants, lead to soil erosion.

It is estimated that invasive alien trees use up around 1.5 billion more litres of rainwater a year than would natural vegetation.[9] That's enough to provide about 12 million people with domestic water for a year, or irrigate 120,000 hectares of croplands. Cape Town and Port Elizabeth are the two cities whose water supplies have been most imperilled by invasive tree species. Clearing the invasive plants in the various catchments could increase the run-off into dams by as much as 55 million litres *a day* in the two metropoles.[10]

If all this sounds bad to you, you can be lucky that, if you live in Africa, there are still grand panoramas where wild creatures roam freely. Not everyone is so lucky. In her profound observation of England's vanishing wilds, Isabella Tree[11] describes a landscape

---

[7] *Ibid*

[8] www.sanbi.org/resources/infobases/invasive-alien-plant-alert/

[9] https://www.hortgro.co.za/news/impacts-and-control-of-invasive-alien-plants-in-south-africa/ (accessed 20 September 2020)

[10] https://www.iol.co.za/saturday-star/news/sas-biodiversity-is-at-risk-despite-commendable-conservation-laws-34745670 (accessed 20 September 2020) (my emphasis)

[11] *Wilding: The Return of Nature to an English Farm* by Isabella Tree (Picador/

where just about everything has been reduced to neat walled fields enclosing "happy" cows and sheep. But the animals are not happy; they are being farmed mechanically. Those farmers who have tried to farm regeneratively are being drowned by the tidal wave of industrial agriculture.

But it's even worse than that. England today is a country so out of touch with wilderness, most people there think "natural" is a tidy, wall-defined farm. The entire country, save for a few meagre tracts, has become a human-altered landscape where few indigenous organisms can, or indeed do, survive. Saddest of all are the songbirds – larks, nightingales, turtledoves – are gone, all gone (to misquote Gerard Manley Hopkins).[12]

In the somewhat smaller "good news" column are the various biodiversity programmes such as Working for Water and Working on Fire, overseen by South African National Parks and SANBI, which have substantial budgets. They underwrite hundreds of thousands of jobs around the country, primarily in clearing invasive vegetation in protected catchment areas.

THE ATMOSPHERE

OUR ATMOSPHERE IS COMPOSED MOSTLY of nitrogen (N_7), nearly 80 per cent, with diminishing amounts of other elemental gases including oxygen (O_8), argon (Ar_{18}), neon (Ne_{10}) and krypton (Kr_{36}). Strange that helium (He_2), the second most abundant element in the known universe after hydrogen (H_1), constitutes only about 0.00002 per cent of the air we breathe.

BUT WE FOUND OUT, BACK in the 1970s, that some gases even less abundant, like ozone (O_3), which occurs only in small traces in our atmosphere (less than 10 parts per million) can be very important for us. Even in that minute concentration, the ozone absorbs around 98 per cent of the Sun's medium-frequency ultraviolet light and thus plays a vital role for all living things on our planet.

We became aware of ozone only when it was discovered that chlorofluorocarbons (CFCs), being sprayed from things like underarm

---

Pan Macmillan, London, 2018)
[12] Misquoted from the poem *Binsey Poplars*.

deodorant and hairspray cans, or escaping from refrigerator pipes, were eating a hole in the Earth's ozone layer – our natural sunscreen. The "hole in the ozone" was most noticeable over the Antarctic, and we realised that, if we didn't reverse the process soon, that hole would get bigger, then we'd all fry to crisps. So, we stopped. That last fact gives hope and just might be an indication that, after all, we are smart enough not to self-destruct.

What it made us realise, those among us who are not already chemists, is that tiny amounts of things can be really beneficial, or equally harmful, depending. Take carbon dioxide (CO_2), for example. As measured in April 2019, this composite gas formed just 0.04 per cent of our atmosphere, but that amount was near the upper limit of being "just right": a few percentage points up or down and we'd be in trouble, plunging us into either an ice age or an inter-glacial heat age. So, it seems we have a small problem here, Houston.

The problem being, over the past 200 years, our burning of fossil fuels – coal, oil and gas – to go shopping or heat our homes, has increased the amount of CO_2 in our skies by a whopping one per cent.

The real issue is that there's no let-up in sight, with many countries (including South Africa) and country-size corporations still heavily invested into producing and using fossil fuels. It recalls the social media meme that challenges us to count our money while holding our breath. And it's accelerating as the human population increases exponentially, gets more and more affluent, wants more cars and more beef and more "climate control" in their homes (not us so much, it's true, in terms of climate control, but we are far from innocent on the other charges.)

The year 2019 was widely reported as having been the hottest year ever recorded, and in the month of July (mid-summer in the northern hemisphere, where there is the most land, which absorbs celestial heat more quickly than does water), being a particular scorcher, along with resultant wild fires.[13] We're sitting on a metaphorical hotplate – and the heat is being turned up.

Images taken by weather satellites during 2019 show a seething planet with vast runaway fires burning across wide fronts in Australia,

---

[13] https://www.unilad.co.uk/news/this-year-was-the-hottest-summer-ever-in-northern-hemisphere (accessed 20 September 2020)

the Amazon forest, Indonesia, Alaska and Siberia, Portugal and California, millions of burning hectares sending billowing plumes of carbon gas into the atmosphere – assisted at the time by a few ginormous volcanoes.

According to the World Meteorological Organisation, the last time this much carbon was released into the atmosphere was between three and four million years ago when Earth's average temperature was 2 to 3 degrees Celsius warmer, and sea levels were as much as 20 metres higher.

That was the time when most of the forest that used to cover our region dried up, died back and gave way to grass, fynbos and succulents. This time round, in 2019, we saw the devastating climax of a five-year drought in South Africa, as well as tornadoes… tornadoes, in South Africa![14] The times they sure were a-changin'.

The simple science of it is that increasing levels of carbon dioxide in the atmosphere (along with a few other noxious airs, most of them also products of human industry, such as sulphur-dioxide, carbon-monoxide and methane, the "greenhouse" gases) create a blanket in the sky, absorbing more heat from the sun, and at the same time trapping in much of the heat that reaches the Earth's surface that otherwise would be reflected back out to space.

The best reflective surfaces are the white ones, ice and snow. However, as we learnt back up in the cryosphere, they are all diminishing. This results in less heat being reflected, more heat being absorbed, more heating up of land and oceans, more melting, more heating… This is called a "positive" (as in reinforcing, not necessarily in a good way, for us) feedback loop.

Predictions are that southern Africa (as well as Australia) is going to get hotter and drier than most of the rest of the world, more quickly. Installing rainwater tanks suddenly becomes a must-do for anyone with their eyes on property resale values.

Remote imagery from NASA shows pollution hotspots around the world in concentric rings from yellow to red, in China, Russia,

---

[14] https://www.businessinsider.co.za/tornado-south-africa-kwa-zulu-natal-severe-thunderstorms-stellenbosch-university-witwatersrand-university-2019-11

India and South Africa among them. The Mpumalanga Highveld, site of 12 huge coal-fired power stations and the massive Sasol Secunda fuel-from-coal plant, is shown in the darkest shade of red.

We are talking carbon dioxide, carbon monoxide and, worst of all, sulphur-dioxide in diabolical quantities. Doctors know this pollution causes or exacerbates asthma, lung diseases, cancers and birth defects. Climatologists know it causes planetary influenza.

The hard science of global warming informs us that we are looking at another 0.5 per cent average increase in the Earth's temperature over the next few decades. If we can stop it there, we could save ourselves, even if we have to adapt to a planet that has more extreme weather events than we are used to, including storms, floods, droughts and rising sea levels. If we don't, and carry on as though it's industrial "business as usual", it's going to be hell on Earth by the end of this century.

But then maybe Donald Trump is right,[15] and the weather boffs are all lying; all 6,000 of the world's top climate scientists who have been tasked with peer-reviewing (picking holes in) the United Nations' Special Report on Climate Change. All of them. It has to be one side or the other that's lying to us: the collective scientific community or the current American president.

The smart people know that climate change is for real, although many are still in denial. Still, that's not the worst of it: Worse is that the majority of people in less developed areas don't even know about it, as was borne out by the Afrobarometer survey referred to earlier.[16] How do you change lifestyles in these conditions?

We have only a very small window of opportunity to reverse the causes of climate change, maybe two generations at most. One thing that is certain is that it's the most important event the human race has ever had to run. On your marks…

---

[15] https://insideclimatenews.org/news/19032020/denial-climate-change-coronavirus-donald-trump
[16] https://www.afrobarometer.org/publications/ad295-are-south-africans-prepared-confront-climate-change (accessed 30 August 2020)

Eco Activists
The Givers and the Takers, Superkids and Groan-Ups

I FIRST MET GIDEON GROENEWALD in the Drakensberg in the 1990s when we were both strapping young men. It was the day I was to begin my solo "grand traverse" of the mountain range. He had walked up to rarefied Witsieshoek Mountain Lodge carrying a shoebox. After introductions, he opened the box to show me the treasures he'd brought. They consisted of some old skull bones and lumps of what looked like squashed balls of clay.

Dinosaur eggs, he informed me. The first ever found in South Africa (or, geologically more correctly, Gondwana).[1] We chatted for some time, looking out across the pleated green ridges below the Amphitheatre, about things geological, palaeontological and generally philosophical. He impressed me as a man of striking physical stature (he looked then, and still does, like he could lead a wagon trek into the wilderness, and you would follow), extremely knowledgeable but disarmingly humble.

He'd heard I was in the area and simply wanted to share his deeper knowledge of the place with me. I wished I could have dallied and chewed the straw with him, and I felt awfully rude to have to cut short our meeting – given how far he'd come to meet me – but I had many

---

[1] "Oom Gideon" was a graduate student under the tutelage of Dr James Kitching (see *The Game Ranger, the Knife, the Lion and the Sheep* by David Bristow, Jacana Media, 2018) at the time. The eggs, of the mammal-like reptile *Massospondylus*, were each about three times larger than a standard hen's egg and were found in the hills near Groenewald's home town of Clarens in the Free State.

miles still to walk that day. Travelling far was, it turned out, a central theme of both our lives.

I had no reason to think I'd ever meet him again, so imagine how chuffed I was to get a Facebook friend request from him some time in 2019. I had no idea how busy he was at the time, but was soon to find out.

Groenewald's name first appeared in the social media when he was part of a Gift of the Givers team that brought water to the stricken town of Makhanda. As has been discussed in chapter 12, these towns were not victims of the drought that had brought many other places to their dusty knees: Rather, it was a mischief created by a – there is no polite way to say this – useless town council, and many believed a crooked one.

But it was events deeper in the Karoo that brought Groenewald to national attention.[2] On a Tuesday morning in early October 2019 a Gift of the Givers convoy pulled into drought stricken Graaff-Reinet. The nation had watched in awe and trepidation as, after several years of below average rainfall, the town's only water source, Nqweba Dam on the Sundays River, slowly dried to a bare clay floor, fly-blown fish rotting on its irregular tiled surface.

The vanguard included a truck carrying bottled water and two tankers filled with water from a borehole they had sunk previously. Passing motorists hooted and waved; people in the filling stations danced on the forecourts. This was hardly their first water-delivery rodeo, but it was the one perhaps most widely reported in the media. "The water is here, the water is here," people shouted. Such was their faith, their hopes, their fears.

It all looked very feel-good, as if it had been scripted, but that was only the overture. All across the country rivers stagnated, dams evaporated, boreholes gurgled. Towns stricken by financial meltdown, those that had failed to take care of their pipes, looked at

---

[2] This section of the chapter was informed by an article by Estelle Ellis entitled "'Oom Gideon' finds 102,000 litres of water in Graaff-Reinet", published in *Daily Maverick* on 11 October 2019. https://www.dailymaverick.co.za/article/2019-10-11-oom-gideon-finds-102000-litres-of-water-in-graaff-reinet/ (accessed 30 August 2020)

their taps and hoped. (In Cradock it was found the water department had not checked the town's supply pipes for at least five years; they were simply blocked.)

But the drought was all part of God's plan, believed "Oom Water", as Groenewald is often called, and he just an instrument of the divine comedy. And then, as if by prophesy, one by one the towns started to go dry – Bedford, Makhanda, Adelaide, Elliot, Beaufort West, Fraserburg, Kimberley. Gift of the Givers stepped in around the country to deliver water in drought-stricken areas where, it seemed to those of us looking on, DWS was unable to do so itself.

On the approach to Graaff-Reinet the carcasses of lambs lay next to barbed-wire fences, the veld beyond chalky, dusty and parched. Across the Karoo hundreds of thousands of sheep died of thirst and hunger. A white bakkie pulled up, and out jumped a scraggy, hirsute man in his trademark leather hat. Groenewald had received the call the night before and had driven through the night from Pretoria. It was his 64th birthday.

He greeted project manager Ali Sablay, and then, after addressing and re-assuring the gathered crowd – "we will find water," looking like an Old Testament prophet – announced that he needed to sleep before the drilling team arrived. With emergency water delivered, their real work would soon begin: securing underground water.

Groenewald is well qualified in geology, hydrology and palaeontology, but he is no *waterwyser* (water diviner), he insists. Using remote sensing technology and his knowledge of the Karoo – he reckons that he and his late wife walked some 40,000 kilometres across the region searching for dinosaur bones – he can locate likely formations. In the dark hours he pores over maps and satellite images; he measures them in 10-centimetre gradations. "Then I pray to God and walk the land," under that distinctive hat, taking magnetic readings every metre.

Most people think it is Oom Water who finds the water but, he maintains, that is the calling of his long-time collaborator Petrus Mofokeng. He merely finds likely spots. They, along with Oom Tienie Landman, have worked together since around the time I first met the water bringer. Doctorates are easy to come by, laughs off the man with the sweat-stained hat, pioneer beard and three post-graduate degrees, but the title of *Oom* you have to earn.

When drilling a borehole, the team knows they have hit pay dirt when Mofokeng begins stamping his feet. The rest see just dust, but Mofokeng sees water, smiles Groenewald. Then the elderly Sotho seer will throw his arms in the air and dance around the borehole like a man half his age. "He is never wrong."

At that same time, the Givers had teams drilling for water in Sutherland, Lusikisiki, Harrismith and Cape Town. In 2019, they hit water 600 times (and twice had their equipment stolen). Sometime later, the drilling team arrived, towing the tatty old caravan in which they live while on the job. The men were tired and hungry; they'd had trouble finding food on the way, as well as drinking water. Within three days they had established several strong boreholes; the town would be saved.

At the centre of this life-giving tableau sits Dr Imtiaz Sooliman, born in Potchefstroom and educated in Durban. After receiving a message from his spiritual advisor in Istanbul, the medical doctor quit his practice in Pietermaritzburg to establish a disaster-relief foundation he called Waqful Waqifin – gift of the givers.

The Givers have establishment hospitals and clinics, created agricultural schemes, dug wells, built houses, developed and manufactured emergency foods, renovated fishing boats, provided scholarships, food and shelter to millions around the world including in Syria, Nepal, Bosnia, Pakistan, Somalia, Haiti, Zimbabwe and in Sooliman's home country.

Groenewald's own journey with Gift of the Givers began in 2014, just days after his wife suddenly died. The phone rang, and a woman who identified herself as working for the foundation asked: "Are you Oom Gideon Waterman?"

Since then, Groenewald and Sooliman have developed a deep relationship, praying together every day for guidance. Groenewald says that, although he is a Christian and Imtiaz a Muslim, their faith is the same, and the lesson is the same: to love thy neighbour.

In early 2020, rains across most of the country broke or at least eased the drought situation. But with the entire planet in the grip of a climate crisis, the big dry will return. It's similar to the sudden outbreak of a viral pandemic: You know there is always something cooking in the microbe-sphere, you just don't know when or where it will hit next. There is much work to do, to ensure we don't get caught again with our trousers down and our pipes blocked.

There are in fact so many "wilderness warriors" out there, like the people who are cleaning up their sections of local rivers, or the citizens of Harrismith who took it upon themselves to fix up their town's infrastructure, as well as all the people working in the bush for rhino, leopard, lion, caracal, pangolin, vulture conservation and more, it's hard to single out individuals.

There are those who work for rhino orphanages and anti-poaching operations, anti-nuclear activists and those who join mass movements like the people – mainly school goers – who marched in Johannesburg, Cape Town and other cities in the great climate change protest organised by the African Climate Alliance in September 2019. These marchers were inspired by the worldwide movement of Greta Thunberg and other schoolchildren participating in strikes and protests to campaign for climate change.

Young Ayakha Melithafa from Eerste River in Cape Town was one of 16 children, including Thunberg, hailing from different countries, who lodged a complaint to the United Nations Convention on the Rights of the Child for not doing all it could to tackle the climate problem.

Melithafa had felt first-hand the effects of the national water shortage when the food garden her mother had planted on a small piece of land failed. This threatened the income of their family and the future of its small farm. So, instead of going to New York, Ayakha tried to organise a mass litter clean-up in Khayelitsha. Only three other people turned up. "We do need more people of colour in the fight against climate change," she told the *Daily Maverick*. "The people that are often affected the most often don't cause the problem."[3]

And then there are people like Lewis Pugh, who swims in the

---

[3] https://www.dailymaverick.co.za/article/2019-09-26-cape-town-teen-climate-activist-aykkha-melithafa-takes-drought-to-the-un/ (accessed 30 August 2020)

world's frozen waters to highlight the plight of our fast-melting polar regions. Or Carla Geyser, who runs the non-profit Blue Sky Society Trust and created the Journeys With Purpose: The Rise of the Matriarch expedition across four southern African countries to highlight the plight of wildlife.

There are the super runners and paddlers David Grier, Braam Malherbe and Peter van Kets, cyclist-turned-ultra-kayaker Riaan Manser and uber explorer Mike Horn, all out there come rain and shine, ice and snow, ocean storms and mountain avalanches, each carrying a flag for a cause. We can't all be supermen and superwomen, but we can each do something.

Take Kolby Harmse, who was seven going on eight when this was written. He was six when he watched the Leon Schuster/Gray Hofmeyr tragi-comedy *Frank & Fearless*, about a mother rhino that is killed by poachers. This spurred him to watch the much more real and gory poaching documentary *Stroop: Journey into the Rhino Horn War*. With his birthday coming round in November, Harmse made a life-changing – possibly life-making – decision.

For his birthday, he asked his people not to give him anything but rather make a donation to rhino conservation. The success of that campaign led to setting up (with a little help from his mom) his own BackaBuddy crowdfunding campaign for rhino conservation. His goal was to raise R50,000 for a Rhino Orphanage in Limpopo. As this book went to print, his donation thermometer on the BackaBuddy website had a temperature just north of R30,000.[4]

She was not quite yet a teenager when Fish Hoek schoolgirl and surfer Jade Bothma began her own environmental movement and non-profit organisation Oceano Reddentes (the ocean revives). Much like Harmse, it was while watching a documentary, titled *Chasing Coral*, that Bothma's own superpowers were awakened: What could she do to help her beloved ocean? And thus was born her mission of "saving the sea one piece of plastic at a time".[5]

---

[4] https://goodnewsdaily.co.za/2019/11/15/seven-year-old-dedicates-birthday-to-saving-rhino-orphans/; https://www.backabuddy.co.za/champion/project/help-a-little-buddy-back-a-rhino (accessed 30 August 2020)

[5] https://www.oceanoreddentes.org/ (accessed 30 August 2020)

Bothma started with organising beach clean-ups, but that just led to more questions: What happens to all the plastic once amassed? Fill plastic cooldrink bottles to make eco-bricks, of course! When she and her fellow "plastic warriors" have made up enough eco-bricks, they plan to build a house for some needy family in one of the poor settlements of the Cape Peninsula. For starters, there's a lot of plastic out there.

Yola Mgogwana was just 10 when she witnessed the event that sparked her environmental consciousness: She and some friends found a dog trapped inside a plastic bag at a flooded illegal dump at Site B, Khayelitsha. It's a tough life there for a kid, but for a dog it can be much worse. (Recall the old saying that, if you have been very bad in a past life, you come back as a village dog.)

She and her friends became volunteers with Earthchild, an eco-organisation that reaches out to children at school, in Mgogwana's case Yomelela Primary. In early 2019, she was among the thousands of learners who marched on Parliament to appeal for environmental justice and action. Not only that, but she was among the few selected to address the crowd on the day.

"I take my eco-activism seriously and spend my spare time raising awareness around the problem of single-use plastic," she told Green Route ZA, a website that describes itself as "an organic network of eco-conscious people working towards making the world a better place". Mgogwana and her friends collect plastic litter, give talks at schools, and monitor the electricity and water usage at their own school, where they have also started an organic food garden.

Already, she says, there are noticeable improvements in the area. But, still young in heart as in years, she says she is happiest gardening, playing soccer or just lying on a patch of grass somewhere.[6]

This is all really just the A, B and C of good neighbourliness: You'd think everyone would get it, and do it – but they don't. Another young warrior who has done something is Aaniyah Omardien who founded the Beach Co-Op action group in Muizenberg and also teaches underprivileged learners for the Waves for Change project.

Her mom was an environmentalist by nature, she says, who

---

[6] https://mygreenroute.wordpress.com/2020/01/08/eco-activists-in-south-africa-yola-mgogwana-earthchild-project-green-route/ (accessed 30 August 2020)

fostered a love of nature in all her children. They were taught to re-use domestic grey water to feed their garden, as well as recycle – and never waste food. "She took us to the beach often, and Dalebrook tidal pool was a favourite summer picnic destination for our family."[7] Now Omardien spends her free time keeping Dalebrook – and other spots along the coast – clean.

If ever you thought one person could not make a difference, take the example of Hout Bay pizza supremo Massimo – he's so good, he goes by just the one name. His has been voted by Top 50 Pizza as the best pizzeria not just in his own valley but the whole of Africa. For every pizza or pasta ordered, he donates a percentage to a local charity.

When it comes to green matters, insalata aside, he serves water (fresh and chilled) in re-usable glass bottles. In a typical month, he says he sells 800 to 900 bottles. A quick consultation with Google suggests there are around 17,000 restaurants in the country that, when you do the sums, add up to maybe 14 million bottles a month. Although, to be honest, not all of the other places are as busy as Massimo's.

Still, not to spoil a good story, in a year he reckons restaurants around South Africa are dispensing as many as 170 million bottles of water – most of them non-reusable plastic. Now imagine if they all did what Massimo does.

It's such common sense you'd think all the grown-ups (what my daughter calls groan-ups) would get it.

---

[7] https://www.dailymaverick.co.za/article/2019-10-22-meet-your-everyday-eco-heroes-who-walk-their-talk/ (accessed 10 September)

Soil and Trees
Blood of the Earth and Breath of Heaven

It was mountain man John Muir[1] who made the observation that, if you pull at one strand of nature, you find you are tugging at the entire biosphere.

Even the shallowest observation will reveal that rocks are the base material of all land and that, decayed, they are also the source of our soils. You cannot easily plant a garden on a rock. Little grows there but lichens and some mosses, a tuft or two of grass or a wildflower in a little cup of soil, maybe a tree like a hardy rock fig that has taken root in some crevice where muck and water collect.

It is soil that is the font of all terrestrial life, namely plants. If the place you inhabit happens to have good soil, chances are better than ever that yours will be a prosperous and easy life. But run aground on barren soil, and yours will be a life of toil and scarcity (unless you hit oil). And from there on, it all comes down to those little capsules of super-concentrated life we call seeds.

A school in my neighbourhood has its ethos based on seeds, acorns specifically, and the splendid oak trees they produce if well nurtured.

---

[1] John Muir (1838–1914) was a Scottish-born American naturalist, also known as John of the Mountains and Father of the National Parks. Following an industrial accident, he walked from Indianapolis to Florida, from where he took a steamer to California (while trying to follow in the footsteps of Von Humboldt). There he virtually discovered the Yosemite Valley for other Americans. It was he who persuaded President Theodore Roosevelt to create the national parks system. He is also the founder of the influential Sierra Club.

The centrepiece of the school grounds is a great, spreading oak that was planted by the first owner of the property, Thomas Bowler, who was for a time an art teacher and went on to become one of the Cape's most renowned landscape artists.

My own school was of very much more recent provenance, but something seen there might have planted a seed in the garden that was my little developing mind. It was a bumper sticker on a car that often appeared ahead of ours on the way to school: an advertisement for a local nursery product, and it read, "I dig super soil". I pondered its various levels of meaning for many years, and in fact still delight in it, the first pun I got.

Some years ago, on a flight from Port Elizabeth to Durban, as we went over the Amatola Mountains and then the Transkei interior, I was shocked to see in such vivid hue the reality of erosion at work. It was a place where too many people were trying to run herds of cattle, as well as goats, on too small parcels of land. Deep dongas tore through the green, the flow patterns all headed seawards.

In the ancient lore of that part of the land, soil, specifically red clay, is known as the "blood of the Earth". The iron-rich clay, or haematite, has been used in sacred rituals since time immemorial and still is. It is thought that mining for the mineral was the precursor to firing clay pottery, making charcoal, and the discovery and smelting of iron and gold in southern Africa. The name haematite means "blood stone". Below me the land had been torn open, and the Earth's vital substance was spewing out across the beloved country.

The greater Ciskei-Transkei region suffered far more than most the ravages of racial division and exclusion. Following the Eighth Frontier War, the infamous cattle killings and subsequent starvation in the mid-1850s, the area became the first coerced labour pool for white-owned farms, mines and other industry: The human machinery that powered the development of this country. It was left to the women and children to plant the mealies and tend the cattle.

Recalling this induced a deep memory to well up. One time, while still free of domestic constraints, I found myself with some mates on the northern KwaZulu-Natal coast, where we had gone for a bit of campin', snorkellin' and spearfishin', when a passing hurricane put us in storm lockdown inside a small tent. The tent held out, but we resorted to smoking the local produce to pass time.

When the storm had passed, we went walkabout in the dune forest. Imagine how surprised we were to come upon a small caravan pitched in an opening, with an elderly woman outside tending a garden. With nothing much else to do, we lingered.

She told us that she and the man we heard typing away inside the caravan moved around, and when they found a nice place, they would pitch camp. He would write while she planted gardens for the local people to enjoy later. When he decided it was time to move on, she bade her vegetal children farewell and moved on to the next place and the next garden.

She also told us that she could see the aura of vegetables and fruit in shops, and how the poisons on them glowed in fierce reds and purples. It was very fairy tale-like, out there in a wild forest and, given our hazy state of mind, quite surreal. Sometimes I wonder if it happened at all. If it did, I could kick myself for not finding out what the "man inside" was writing, and who he was.

In subsequent soil diversions I have come across the likes of Australian soil ecologist and evangelist Christine Jones. For the past several decades she has travelled the world (including South Africa), giving talks and workshops to other ecologists, farmers and gardeners on subjects ranging from biodiversity and carbon sequestration to water management and restorative land management.

The example she has set has spurred others on to take leaps of faith in life, like Capetonian Pat Featherstone. She was born in Zimbabwe when it was still a colony, playing in the veld, but preferably the mud. Her mother sewed hardwearing outfits from food sacks for her and her sister, so they would not soil their good clothes while mucking about. Part of that was working in their mother's food garden.

Then life happened, and she went to university, completing her BSc honours in Harare. She moved to South Africa and started teaching. But, by and by, she got fed-up with it and soon found herself "floating about" – until she remembered her mother's and grandmother's food gardens and fruit trees and her own childhood ramblings. That led her to the Food Gardens Foundation, an organisation that sprang up in

1976 to bring food gardening expertise into beleaguered townships.

Ten years of working there, and Pat saw an opportunity to start her own soil-based operation. With its HQ down a bridle lane in Constantia, Soil for Life works mainly with home gardeners on the Cape Flats[2] and aims to bring food security to poor communities all across South Africa.[3] Other such organisations include Food & Trees for Africa and Abalimi Bezekhaya (farmers of the home).

Turns out that – surprise! – even the world's number one futurist, Elon Musk, is something of a tree hugger. In a Twitter conversation in early 2019 he promised to sponsor a million trees to the #TeamTrees initiative, which was aiming to raise US$20 million to plant 20 million trees.

But arguably South Africa's first soil and tree champion came from much less auspicious origins. When I announced to my family that I had decided to become an environmentalist, my mother (far more inclined to clasp gold than hug a tree) gave me a fridge magnet. It is still there long after she is gone. It reads: "After 2,000 years of civilization how can you make a road and not plant trees on either side? Robert Mazibuko." He sounded like my kind of guy, so I endeavoured to learn more about him.[4]

Robert Mazibuko was born around 1910, the fourth child of eight to parents who were labourers on a farm near Spionkop in the Natal thorn belt. His and the other working families on the farm were allowed to cultivate small acreages and run some stock, as well as enjoy a share of the larger farm's harvests. All the children tended the family fields as well as those of the white farmer, learning all the tricks and trades of the land.

At a mission school on the farm, they learnt how to build traditional huts and to make things like clay pots, grass mats and milk strainers, as well as wooden utensils and farm implements.

---

[2] In Cape Town all the old monied suburbs hug the fertile slopes of Table Mountain, while the working class and townships sprawl out across the sandy and wind-blasted Cape Flats.
[3] https://soilforlife.co.za/ (accessed 30 August 2020)
[4] Biographical information on Robert Mazibuko is taken from *The Tree Man: Robert Mazibuko's Story* by Joanne Bloch (New Readers Publishers, Durban, 1996).

This allowed Mazibuko's family to accumulate some capital, which afforded them the luxury of sending him (alone among his siblings) to attend school beyond Grade 4.

He showed such promise that, on finishing Grade 7, he was sent up to the prestigious black-run Driefontein Secondary School. One of the subjects taught there was gardening. Once again, the lad from Spionkop excelled, and so was sent on to the St Francis Teachers' Training College, part of the Catholic seminary at Mariannhill. There he was taken under the wing of a Father Huss, who taught agriculture.

Mazibuko enjoyed a head start over most of the other students and was soon put in charge of all the school's gardens and trees, and they in turn flourished under his green fingers. Education inspectors would come from as far as Pietermaritzburg, the provincial capital, to admire his handiwork. Beyond gardening the students had to fix windows, doors, desks, whatever needed to be done, even building toilets in the surrounding communities.

It was there, following the organic growing methods taught at the college, that the young man saw and understood the interconnectedness between bees and other insects, birds and plants. Upon graduating in 1930, the young gardener went on to work at mission schools around Natal. He taught others how to grow vegetables and flowers, plant and tend fruit trees and – most important of all – how to make healthy soil: "black gold", he called it.

After a quarter of a century of school teaching, he joined the Valley Trust, an agricultural and community health project in the Valley of a Thousand Hills. There he refined the trench gardening system he'd been developing all those years. Trenching is specifically suited for places with poor soils and/or low rainfall: Very simply, you dig a trench, fill it with organic matter and some soil. Water it, and the trenches become rich, moisture-retaining micro-habitat valleys.

Nearing his eighth decade, the man who had become known as the Tree Man of KwaZulu-Natal procured overseas funding to start his own place. This was the Africa Tree Centre, in the heart of Zululand, where he continued his teaching. Finally, he was able to blend the many influences of his life, from simple farming to what can be considered "modern" organic agriculture, the African bond with the earth and traditional medicine, as well as the spirit of ubuntu, or sharing, that reflected both his traditional as well as his missionary legacies. Mazibuko died in 1994, having spent his entire life observing and working with natural processes.

No one better understood John Muir's concept of the connectedness of all things in nature. How trees breathed in air through their leaves to make molecular skeletons around a carbon core, and from sunshine to make sugars. From the soil they sucked up water with dissolved minerals, including nitrogen, phosphorous, potassium, calcium, and silicon oxide, which gives them their flexible rigidity.

Soil and trees, marrying the breath of heaven with the blood of the Earth, are among the greatest behests of Mother Nature.

Geological and Environmental Health

Why Where *You Were Born and* How *You Were Born Can Really Matter*

THE FIRST QUESTION THAT SPRINGS to mind is, is geological and environmental health really a thing? Historically (at least in the sense of modern science), geology and medicine have been regarded as quite discrete disciplines. Yet, as Hippocrates appreciated, there is significant overlapping.

Over time, people and animals have learnt by trial and error what to eat and what to avoid in their areas. Like white snakeroot on the Great Plains of North America: The plant, *Ageratina altissima*, contains the deadly compound tremetol.

Abraham Lincoln would have known not to eat its lacelike white flowers, as would his mother, Nancy Hanks Lincoln. No one was unduly alarmed when Mrs Lincoln, having recently drunk cow's milk, began losing her appetite, then became nauseous and weak, since these were common maladies back in the days before refrigeration and sterilisation of foods. Even when she developed stomach cramps, no one worried unduly. But then her tongue turned bright red, and she said it felt like her blood was boiling inside her. She died soon afterwards. She had merely drunk the milk from a cow that had grazed on snakeroot in a pasture.

All along the sub-tropical eastern margin of South Africa, people would have known not to eat the fruits of the forest poison rope, *Strophanthus speciosus*. The capsules contain a deadly cardiac glycoside called strophantin that has been used as an arrow poison.

Unfortunately, this was not something the straggling castaways of the *Grosvenor*, which wrecked on the Wild Coast in 1782, would have known. By the time they reached Brazen Head (about 50 kilometres south of the wreck site), fewer than half the survivors were still alive, and they were already starving. Several who sampled the fruits of the deadly vine died right there.

Plants get many of their basic elements and compounds from the soils in which they grow. On his journeys along the Silk Road in the 1270s, Marco Polo noted that the party's horses became distressed in the mountains of central Asia: Some of the horses lost their hooves and had to be dispatched. We now know that it was selenium poisoning from plants they'd eaten.

Farmers know selenium poisoning as the blind staggers that can affect any grazing stock. In South Africa the chief culprits are the delicate yellow daisy-like flowers of the various species of *Senecio*, commonly known as ragwort, poisonous ragwort or grasshopper bush. The genus is one of the largest groups of flowering plants, with some 1,200 species occurring mostly in Mediterranean regions and temperate montane grasslands.

If you are not by now convinced that the correct answer to the question posed above is a positive, I have a 3.5 kilogram, 777-page book that confirms it.[1] The tome expounds upon the subject in detail, as you would expect of a major medical reference. Among the bewildering array of subjects covered are geology and its effects on our biology, extreme occurrences of minerals in the environment, volcanoes, arsenic in groundwater, concentrations of heavy metals and, of special interest to us, the impacts of mining on our general health.

My favourite chapter is about "geophagy", and it's one most parents will identify with: when your toddler ingests soil, ants – and worse. But humans have been eating earth, or soil, for millennia. From the time of its first description in ancient Greece to the end of the 16th century, one of the most popular medicines was Lemnian

---

[1] The big book in question is *Essentials of Medical Geology* (revised edition) by Selinus et al (Elsevier/Springer, Dordrecht, New York, London, 2005, 2013).

earth, or *Terra sigillata*. The original concoction was made from soil from the island of Lemnos, mixed with the blood of a sacrificed goat, compressed into a tablet and stamped with the image of a goat.

There are numerous accounts of when people, entire nations in some cases, have been driven to eating soil (among many unsavoury things) through extreme starvation. In such times, drinking one's own urine has also been a fairly common practice. In both North and South America, slaves sometimes committed suicide by eating soil, sometimes voluntarily to escape hell on Earth, but also sometimes through sheer hunger.

Some old tribes have long practised geophagy, mostly of clays, as an aspect of folk medicine. They might have learnt it from primates that regularly eat clay to combat intestinal parasites. Even parrots are known to eat clay in the wild: some of the nuts they eat contain toxins, so it might be to neutralise these.

The list of animals known to practise geophagy is long and includes both Tabby and Spot (like when they swallow grass to help void something they've eaten), and livestock that are given mineral "licks" as dietary supplements. Even elephants need mineral supplements and in some places actually mine salt deposits. There are caves on Mount Elgon in Kenya that have been excavated by tuskers for this purpose. The old safari clanger "do giraffes hunt in packs?" has its origins in the towering camel-leopards chewing old bones for their calcium.

There is some evidence that eating soil and generally mucking about in mud is not just good for us but even essential. There is an idea being bandied around medical corridors that as human societies grow increasingly distanced from their rural roots, allergies seem to be increasing proportionately. Our immune systems need contact with Mother Nature in order to develop fully. The apparent allergy pandemic in urban kids is also often attributed to all the bad stuff being put into mass-produced foods. Some say it's the pesticides and herbicides with which crops are sprayed. It's most likely a combination of all of the above.

But don't panic, keep calm and carry on, because we are all going to die. The only things that need to be negotiated are when and how. That could well depend on where you were born (farm or flat) and how (natural or by Caesarean section). All other things being equal, the best scenario health-wise would be on a farm and by natural childbirth. The opposite would be by Caesarean section delivery. These factors could also determine how you fare in life in the years between.[2]

We each have inside of us around 38 trillion microbes, which we acquire in the first five or six years of life. Naturally born children are gifted a very large number in their mother's birth canal and are thus buttressed against many maladies awaiting us in the big bad world. The rest we get along the way, but children with close contact to nature get theirs much earlier, and thus their immune systems are more robust much earlier. Eating soil – and even dog poo – is all part of that process.

Medical studies have gone even further, suggesting at least some part of our personalities are also formed in those early years, through the assemblage of tiny organisms with which we come into contact during this time. If, like the avenging Macduff in the play *Macbeth*, you were not of woman born, you might have been disadvantaged from the moment you made your grand entrance.

Speaking of children, the story of Hansel and Gretel by the Brothers Grimm has delighted youngsters for many generations. It recounts the travails of two children of a poor German woodcutter and a nasty stepmother. They get lost in a forest but manage to outwit an old witch who lives in a gingerbread house and had planned to eat them, in short. The backstory is there was a famine in the land, so the horrible stepmother takes the siblings into the woods and leaves them there to starve, so she won't have to feed them.

Originally Jacob and Wilhelm Grimm did not set out to write children's stories, but rather to preserve German folklore. The story in question was originally called "Little Brother and Little Sister", and it was set, quite specifically, between 1315 and 1322. At that time, Europe was in the grim vice of a terrible famine, precipitated by extensive volcanic eruptions in the west Pacific Ocean.

In the first year of the famine, mostly only the poor serfs suffered

---

[2] This section is largely based on *The Body: A Guide for Occupants* by Bill Bryson (Transworld/Penguin Random House, London, 2019).

the hunger of failed harvests. However, as the years dragged on and the famine worsened, landowners and eventually even the squires and ladies resorted to eating their leather shoes, their children or one another. Some parents starved themselves so their children might survive. Others ate the children so they might. It was the worst food crisis of the Medieval Period in Europe, and millions died all the way from Dublin to Moscow. Just a quarter century later, the region was smitten by the Black Plague.

Much more recently, when the volcanic peak of Pinatubo erupted in the Philippines in 1991, the aerosols ejected into the upper atmosphere influenced the global climate for years thereafter. Geologists estimated the eruption dispersed around 800,000 tonnes of zinc, 600,000 tonnes of copper, 550,000 tonnes of chromium and 30,000 tonnes of nickel from the bowels of our planet into the biosphere.[3]

Along with those generally beneficial elements came a much less welcome load of around 1 million tonnes of lead, 10,000 tonnes of arsenic, 1,000 tonnes of cadmium and 800 tonnes of mercury, as well as smaller amounts of even more dangerous stuff such as beryllium, radon and uranium.[4]

At any given time, 50 or more volcanoes are erupting around the world bringing a lucky packet of minerals and weather, good and bad.

Many South Africans who grew up in the precious-metal-laden regions of Gauteng, Mpumalanga and the North West might very well have experienced the side-effects of mining. The most obvious is air pollution from coal mining and coal-fired power stations, along with the chronically noxious slime dams that still surround many of our old gold mines: a failure there, and it's zombie apocalypse time for anyone caught downslope.

For the miners themselves the stakes have always been the highest. The number-one killer underground has always been silicosis: the inhalation of rock dust from blasting and drilling that kills lung tissue much the same as cancer. Up to the 19th century, it was thought to be tuberculosis, which was one of the great killers worldwide at the time and in some places still is.

---

[3] *Essentials of Medical Geology* (see footnote 1 in this chapter).
[4] *Ibid.*

Down the ages silicosis is thought to have accounted for the deaths of many millions of miners, ever since the time large-scale mining began in ancient Greece and Rome (and possibly even before that in the Levant).

ONE OF THE MOST POIGNANT works in the oeuvre of Afrikaans poetry is 'In Die Hoëveld' by Toon van den Heever. It tells of an ailing miner sitting inside the dark tunnel of a gold mine and dreaming of his old farmstead "that he had to leave for money", a place where the grasses dance in the wind and antelope tip-toe among the headstones of the family graveyard. It's a place where a person could still believe in a God, the poem concludes.

It hardly needs saying the mining company Randlords, living it up in their mansions on the Parktown Ridge or in Britain – then as much as now – have never been keen to accept any responsibility for the working conditions of the actual miners, or the consequences.

However, following a class action taken out in 2012 against six of the largest gold mining companies, a settlement of R5 billion was finally reached early in 2020, and anyone who had contracted the dreaded silicosis while working at one of these mines is entitled to claim.[5]

First hurdle over, the next will be to locate all those affected, mainly migrant labourers from the Eastern Cape, Lesotho and Mozambique, and somewhat fewer from Zimbabwe and Malawi. The number could be 1 million people since 1965, which was the cut-off date set by the court. (Presumably anyone working in the mines before then who had contracted the disease was considered to be already dead.) The final hurdle will be actually getting the money paid to them. *Shosholoza* – "into the mountains the steam train is running" – as the old working song goes. At the time of going to print, no payments had actually been forthcoming.[6]

Maybe the greatest disdain for workers ever shown by the mining

---

[5] https://www.dailymaverick.co.za/article/2020-02-06-first-payments-to-claimants-in-historic-silicosis-settlement-set-for-second-quarter-of-2020/ (accessed 11 September 2020)

[6] *Ibid*

industry occurred in Kimberley in the late 19th century. If you worked there, you worked for "the man", and that man was Cecil John Rhodes, one of the richest and most powerful men in the British Empire at the time.

Starting in 1882, a smallpox epidemic swept across the Cape Province. When it became known that it had reached the diamond capital of the world, Rhodes's equally infamous sidekick Leander Starr Jameson, a medical doctor, was tasked with the job of preventing people finding out the truth, fleeing the place and averting work stoppages. Jameson devised a plan, along with several other doctors under the sway of Rhodes, to disseminate declarations that came to be known as the "pink slips" infraction.[7]

Stating the illness was a rare but totally curable kind of skin irritation and, basically, just keep calm and carry on, these slips were delivered and posted all around town by young runners. Finally, Hans Sauer, a local doctor who was not in Rhodes's camp, convinced the powers back in Cape Town of the magnitude of the problem.

Sauer was promptly made medical superintendent of the district and by 1885 had managed to get the epidemic under control. However, by then many thousands of people – mostly black labourers – had died of the killer disease. Sauer thought the number might have been higher.

When it comes to the geo-medical effects of mining, nowhere in this country eclipses Penge (now known as Mabulane), the Chernobyl of South Africa. It's a lovely part of the country, where the swift-running Olifants River emerges from the Wolkberg Wilderness Area, with villages sprinkled all along its banks and up the tributary valleys.

Penge is a ghost town along with some living ghosts: People still eking out a living by harvesting old pieces of metal from the now-closed asbestos mines and appropriating building materials from the decaying structures. Many of them are gravely ill. Penge was never officially evacuated or rehabilitated, as it should have been. About

---

[7] You can read a more comprehensive story of the "pink slips" in the book *Of Warriors, Lovers and Prophets: Unusual Stories from South Africa's Past* by eminent journalist and author superlative Max du Preez (Zebra Press, Cape Town, 2010).

2,000 people still live in and around the old dumps of extremely fine asbestos dust, which winds gather up and spread across the landscape like fresh fallen killer snow.

The particular type of asbestos that was mined at Penge was amosite, or brown asbestos. It causes asbestosis, which has been labelled as "the grand-daddy of occupational health killers".[8] The mine closed in 1992, but two others in the area continued production into the 2000s. Officially, the only people left living here are the few police officers in what used to be the mine surveyor's office. Tacked to a wall is a notice, stuck up next to the mine's once proud brass plaque, exhorting anyone with a health claim to visit certain lawyers' offices in nearby bustling Burgersfort.

Mining operations there began around 1910 when asbestos was considered one of the wonder products of the 20th century, along with plastic, leaded petrol, DDT and thalidomide. Asbestosis is a chronic inflammatory lung disease, often leading to lung cancer and mesothelioma that attacks the linings of the internal organs.

Once the terrible effects became known (and they did as early as the 1960s, but the knowledge was suppressed), the reaction worldwide was near immediate and perhaps over-compensatory. Having asbestos roof tiles or an asbestos radiator would be harmful only if you ate them. But not so for the miners, and slightly less so for the other residents of the mining hamlet. So far about 3,000 claims have been lodged against the two mining companies involved, the Cape Asbestos Company and Tuner & Newall, based in Manchester.[9]

On an even darker note, it was to the abandoned Penge mine that the C10 "special operations" security police unit – under command of the man codenamed Prime Evil, Eugene de Kok – took, tortured, shot and killed ANC operatives.[10]

A final episode on environmental health concerns the extremely Byzantine goings-on at Musina (previously Messina), on the Limpopo

---

[8] https://www.theguardian.com/world/1999/jan/31/davidberesford.theobserver (accessed 20 September 2020)

[9] *Ibid*

[10] In 1996, De Kok was convicted on 89 charges and sentenced to 212 years in prison. He was paroled in 2015.

River. Although designated the country's eighth largest watercourse, the river is mostly seasonal and flows more often with sand than with water. For decades past, intensive farming, and diamond as well as coal mining in the area, have made heavy demands on the limited local water supply in one of the hottest places in South Africa. (This point goes to the heart of the matter, to which we will return.)

It is the proposed location of one of the largest industrial projects ever envisaged for South Africa, of a scale equal to Richards Bay or Coega, but it is one the government tried its damnedest to keep a secret. When amaBhungane investigative journalists got their hands on three bulging lever-arch files detailing the scheme, they struggled to decipher the circular references and legalese.[11]

That was until they came to an – at first seemingly trivial – PowerPoint slide presentation. When observed more closely, however, they realised it spelled out the entire scheme in language and symbols even a primary school learner could understand. And what that scheme was, was a dark and dirty ploy of deceit and intimidation.[12]

The first order of business was getting a lease on the Mulambwane community lands, which would be required for the project. It would take a host of contract lawyers to sort through the Machiavellian connivances in those documents.

Mashudu Samuel Muluadzi, headman of the community at the time of signing the land lease documents, is a smart man, and he did recognise the trap into which he was being lured.

When he did, he took legal action … and was summarily removed from his seat on the tribal authority and replaced by pro-project people. The civil case he has been pursuing since around 2018 has been stalled at the local police station ever since. It is hard to unravel all the paperwork and shenanigans, but to get a grip on this slippery fish, one has to look at its genesis.

In 2011 Dr Pieter du Toit (originally an oil and gas engineer)

---

[11] https://www.dailymaverick.co.za/article/2020-04-01-killing-the-holy-ghost-inside-the-r145bn-plan-that-would-destroy-the-limpopo-river/ (accessed 11 September 2020); https://www.dailymaverick.co.za/article/2020-05-04-how-a-r10-7bn-zero-waste-megaproject-was-buried-by-limpopos-chinese-deal/ (accessed 11 September 2020)

[12] *Ibid*

and his business partner Deshika Kathawaroo, an environmental scientist, pitched the plan of creating the world's first zero-solid-waste eco-industrial zone to then Limpopo premier Cassel Mathale. They named it the Limpopo Eco-Industrial Park.[13] Mathale was impressed.

In setting up Eco-Industrial Solutions (EIS), Du Toit and Kathawaroo aimed to equip public facilities to shoot for a zero-waste industrial dream. All went swimmingly for some time for this revolutionary ecological initiative, with memorandums of agreement and intent (MOUs and MOAs in the jargon of the game) flowing swiftly between the parties.

By March 2014 it looked like the industrial park would be green-lighted by the Department of Trade and Industry. But thereafter, things began to go pear shaped.

In September 2014, the top brass from the Limpopo Economic Development Agency (LEDA) visited the People's Republic of China. There they were introduced to a certain Mr Ning Yat Hoi, chairman of the Hoi Mor Group that would, in time, be named as the anchor tenant of the Musina-Makhado industrial project.

The first indication of the EIS no-waste project's speed-wobble was when, according to Du Toit, they started getting "outrageous" requests for unwarranted paperwork from LEDA. When the final Limpopo Special Economic Zone (SEZ) plan was announced in September 2017, it was clear 126 pages of the original EIS application had been appropriated and given to Hoi to present as his own.[14]

Around that time, it was reported in the *Daily Maverick* and the *Zimbabwean Sunday Mail* that there was strong evidence that Mr Hoi had appropriated several million dollars from London-listed nickel and gold mining corporation ASA Resources Group, operating in southern Africa.

You can only imagine how complicated things got after that, with rivers of denial (from both Hoi and his South African sponsors), lawsuits and counter suits filed by just about everyone, including the Cape Town-based Centre for Environmental Rights, which claimed

---

[13] https://www.limpopoecoindustrialpark.com/ (accessed 20 August 2020)
[14] https://www.dailymaverick.co.za/article/2020-04-07-earthcrimes-part-one-limpopos-dirty-great-white-elephant/

the Limpopo and Chinese developers had contravened various laws and had committed perjury in the process.

This is the point in the story where two big bull elephants stride into the clearing and put down their enormous feet, metaphorically speaking (we have no information about what their shoe sizes are): South African President Cyril Ramaphosa and Chinese Party General Secretary Xi Jinping. In 2018, the two signed a deal to cooperate in the fields of climate change, water resources and transport infrastructure which – it turns out – was reported to be a smokescreen for the Limpopo SEZ project.[15]

By early 2020, nine Chinese companies had committed to investing R145 billion in that far northern, hellishly hot, baobab-speckled valley where a river seldom flows.[16] Now is a good time to remind you, dear reader, that our beloved president is a Limpopo man, and in my view he may see this special economic zone development as his legacy to his people: much in the way that PW Botha did with the Mossgas/PetroSA installation. But the entire shebang could just as easily backfire.

As reported in *Daily Maverick*, the Chinese-funded industrial development node will comprise a 1-million-tonne ferro-manganese plant, a 500,000-tonne silico-manganese plant, a similar sized ferro-silicon one, all contained within a 60 square kilometre fenced area (the disputed community land). Inside there would also be an office park, administrative and living facilities, five-star (naturally) hotel, another three-star hotel for commoners, residential apartments, a shopping mall and good old toy-town-style farmer's market.

But the kicker to all of this is that it will be powered by its very own monstrous coal-fired power station. And to think it all started with an innovative, ecologically sound, climate-sensitive industrial park. The weather scientists might have lit upon a project squib dampener with this. Two of South Africa's leading ecologists, professors Bob

---

[15] https://www.gov.za/speeches/president-cyril-ramaphosa-concludes-state-visit-china-3-sep-2018-0000 (accessed 20 September 2020)
[16] https://www.dailymaverick.co.za/article/2020-04-01-killing-the-holy-ghost-inside-the-r145bn-plan-that-would-destroy-the-limpopo-river/ (accessed 11 September 2020)

Scholes and Francois Engelbrecht, reckon the project will spew the emissions of its own demise.[17]

They point out that by 2050 the Limpopo Valley is predicted to experience temperatures on average about 4 degrees Celsius higher, where it is already oven hot most times. Looking forward, for all but about two months of the year it will suffer continuous heat-wave conditions that will require air-conditioning just to function, at home or in an office – never mind inside those smelting factories – both day and night.

That means the power station would have to work full steam, polluting the air and contributing to global warming. And don't forget to mention the drying river that will have to provide all the necessary water, including cooling for the factories and the power station.

There is a saying in that part of the country that when bull elephants fight, it is the grass that gets trampled. In this case it will be the grass, the trees and all the little people. They might never get to see the Limpopo flowing again.

---

[17] https://www.dailymaverick.co.za/article/2020-04-01-killing-the-holy-ghost-inside-the-r145bn-plan-that-would-destroy-the-limpopo-river/ (accessed 11 September 2020)

Green Living – and Dying
It's a Fine Art

I REMEMBER AN OLD GAME: Would you eat something – typically something gross like a live garden slug, or a fresh dog turd – for R1,000? No? How about R10,000? A similar scenario that, apparently, saw the actual harsh lights of a TV quiz show was "would you take the money?"

You are offered a suitcase stuffed with money, typically a million rands, pounds or dollars. However, if you do take it, it is with the knowledge that somewhere in the world (preferably a faraway, overpopulated place like India or China), a child will die by virtue of the process we know as the butterfly effect. Would *you* take the money?

It brings to mind one of those apocryphal things Winston Churchill is supposed to have said but probably never did; he was too well bred. Anyway, he's at a fancy dinner party when he asks an archetypal aristocratic woman seated next to him (no, this is not the one about he being drunk and she being ugly):

"Madam, would you sleep with me for a million pounds?"

She replies in the affirmative.

"And would you sleep with me for ten pounds?"

"Mr Churchill," she gasps in indignation, "what do you think I am?"

"We've already established that madam, now we are merely haggling over the price."

We all do, we all take the money every day, each time we drive a car, or fly in an aeroplane, turn on a heater or even cook a meal. In the great inter-connectedness of the biosphere, because of this, somewhere a baby dies. It's usually a young chimpanzee or orangutan whose forest home is being chopped down to plant fruit palms to make palm oil, so that your child, or the brat next door, can have the Oreo for which they threw a tantrum in the supermarket.

By the same knock-on effect, we are all complicit in the destruction of nature, a small comfort being that we down here in the Southern Ocean are not as bad as some people. I recall being in Florida in the United States one muggy summer. Just walking around outside brought on a sweat. My partner and I ducked into a shopping mall – where we were blasted out again by the sub-arctic air-conditioning. They just do everything there in extremes, whether it's kindness or craziness.

I remember laughing to myself during the very funny movie *Beverly Hills Cop*. There's a scene where the hobo character played by Nick Nolte goes rummaging in a dustbin for food. It was huge, so big in fact it had wheels! It was a time when we still had old-fashioned metal dustbins in South Africa. Now we too have plastic wheelie bins, four times the volume of the old ones, same as they did in the United States back in the early 1980s. Mostly, they are filled with plastic and thrown-out food. But I'm no longer laughing.

We are each of us that person in David Bowie's song "The Man Who Sold the World" (everyone you meet, face to face, is that person): We sold the world for mostly trivial increments in comfort and convenience, for trinkets made from rhino horn and sea shells. Now's the time to start buying it back, one piece of litter at a time, with all the money we made while selling it.

There are a few good reasons why we need to reassess the way we live. One is that most of us who have First World lifestyles use way too much: too much petrol and too much toothpaste, too much sugar and too much food generally. But more telling is that we use way too much electricity and water. We buy stuff we don't need, then we throw stuff away, much of which we didn't need in the first place. And with that comes all kinds of waste and pollution and other problems down the line, for ourselves as well as for others.

In South Africa we have become captives of a centralised state that has failed us – miserably – with service delivery. Although we might just have bottomed out of the "state capture" pit, it will take untold years before things get back to tolerable levels; maybe longer depending how things go on the political front. This alone is a pretty good reason to go running for the hills.

One of the best environmental lessons we learnt in recent times was during the acute water shortages across South Africa in the late 2010s: that was how to conserve water, mainly by using less. For many of us, it opened our eyes to the fact that clean water is a limited resource, not just something that, on demand, pours out silver spouts in our homes.

A bad habit the COVID-19 lockdown highlighted was commuting. Already in the first week of "stay put", people in various cities around the world posted on Facebook claiming considerable reductions in air pollution. On the first day, people in Pretoria said they could see Johannesburg's jagged skyline clearly for the first time in many years. In the days following, you could virtually hear Mother Earth breathing out and the sounds of nature filtering back in.

MOST WORKING PEOPLE WHO LIVE in cities commute. The highways and arterial routes are all clogged up with people, mostly alone in their car, frothing about the roads being gridlocked. What is most amazing is that, what with people spending up to two, three and even four hours a day in this snail-crawl scenario, there are not many more deaths by road rage. Even in the time of breakdown of our passenger rail service, how many ever stop and ask themselves what kind of madness it is? Surely there must be better ways to do this?[1] E-bikes come to mind.

In our homes, the amount we throw away is criminal. Have you ever visited a big city landfill? Not your local recycling dump where everything is tidily kept in skips, but a mountainous dump-all where the municipal garbage trucks roll in about one a minute, all day, every day, carrying decaying food, old paint tins, soiled nappies, building

---

[1] The COVID pandemic has taught us that most of us can happily work at home and that commuting is a nut's game. Unfortunately, many big businesses don't seem to like it: Essentially, they don't trust you. But they too might have figured all the rental money they could save. I predict the next big trend will be all those office blocks around the world being converted for accommodation.

rubble, plastic, medical waste and worse that has not been disposed of responsibly.

The national water crises coming on top of Eskom's load-shedding fiasco, and then topped by the Corona pandemic gave us the opportunity to pause and reconsider how we live, what most matters to us, what our priorities are, and what we could easily do without (other than toilet paper, booze and tobacco).

When it comes to saving water, there are several ways in which we each can make a difference. The first is obviously to use less: taps running, long showers, deep baths… Baths, like cassette tapes and CDs, will none too soon become things of the dinosaur past. For those with gardens, the next two no-brainers are installing rainwater tanks and grey-water recycling systems.

Then the big one – power, krag, umbane, call it what you will, it's the stuff that fuels empires or sees them fail. For both environmental and political reasons, going solar, if you can afford to, is a very good option to becoming independent and being environmentally responsible. For a new build, going fully 12-volt solar is now a viable economic option, but less so if your home is already plugged into the 220-volt national grid.

The heavy drawers – ovens and stoves, washing and dish-washing machines, heaters and swimming-pool pumps – contribute a large share to a typical suburban home's energy bill. But by far the biggest energy glutton in your home is your hot-water geyser, or geysers (many of them seldom used).

Basically, the geyser is a large kettle that keeps the water inside at a temperature somewhere below boiling point (usually 50 to 60 degrees Celsius). Most sit inside the roof space, which is just about the best place for it to lose its hard-won heat. (Covering it with a thermal blanket is a surprisingly big and easy money-saver.)

Installing a solar geyser will give you the biggest savings and will pay for itself in two to three years. But there is a far easier and very much cheaper way of reducing your geyser's (or geysers') hunger: Turn it, or them, off whenever you don't need piping hot water.

Most households require hot water first thing in the morning and then in the evening. It takes about half an hour in summer and one hour in winter (unless you live in Sutherland or Dordrecht)[2] to get it up to heat. So, all the allocated person needs to do when they get up to make that first morning cuppa is to switch on the geyser. Then, with all the morning ablutions and washing done, simply switch it off again.

If you happen to be the allocated person, the smart thing is to get a sparky to install a self-timing switch at your electrical board and all the switching on and off will be taken care of automatically.

Packing up, scaling down and heading off to live in a small town or on a piece of land, going off the grid, increasingly becomes an aspirational goal. Going counter-cultural might not be within everyone's grasp, or indeed be their post-hippie dream. But whether you are a town mouse or country shrew, we all need to look inwards rather than outwards to find our salvation. There is much we can do to lighten our footprint, become more independent, and cut our expenses and waste.

With anything new, it's always getting started that is the hardest part. Most of our waste emanates from our kitchens, so it would be a good kick-off place. My son once worked out the carbon cost of the foods in his home for a school project. You'll be surprised to learn that what came out on top, in the worst way, was Oreos.

When he researched the ingredients and their likely places of origin, he found the standard "cookie" (as they call them in their place of origin) came from:

- Sugar – Caribbean region, Brazil, Indian Ocean
- Flour – United States/Canada
- Canola oil – Canada or South Africa
- Cocoa – West Africa
- Corn syrup – as for sugar
- Leavening agent (baking soda) – United States
- Corn starch – United States
- Salt – various

---

[2] Sutherland is officially the coldest town in South Africa with an average winter low temperature of only marginally above 0 degrees Celsius. The coldest temperature yet recorded in the country was on Buffelsfontein farm outside Dordrecht, -18.6 degrees Celsius.

- Soya lecithin – USA, Argentina, Brazil, China
- Vanillin – Madagascar, Mexico, Tahiti
- Chocolate   various
- Palm oil – Indonesia and Malaysia

All those products are shipped or flown to the central manufacturing plants, four in the United States and one in Mexico. From there the finished biscuits – the ones that reach us here – are routed via a worldwide distributing centre in Bahrain, from there to central warehouses in South Africa, and finally to the individual stores. That is one heck of a lot of carbon for a little treat.

And don't get me started on palm oil… Then again, do. It is the product that is perhaps the most destructive kind of farming ever foisted on the planet. You should never, ever, buy anything that contains palm oil – although it's hard because the stuff has become ubiquitous in mass-produced confections.

Hopefully by the time you finish reading this book you will have become an ardent food-ingredients label reader. (You might need to take powerful readers with you when you go shopping, since the makers like to print these labels in very small type.) Among the top order of things to avoid are palm oil, hydrogenated fats and anything labelled "low fat" or "added sugar". They're pure poison.

I went to my local supermarket recently to check up on a few things and zoned in on their peanut butter and honey stocks. All the big honey brands stated the source as being South Africa and/or China, sometimes via India or Zambia. That's a sure sign it is wholly, or partly, fake honey.

Next I checked the nuts section – and that was worse. Every one of the many peanut brands contain palm nut oil. Excepting here the food-label spin-doctors have been as busy as a swarm of robber bees. What the labels now say is "vegetable oil (palm fruit)" to try to obscure the fact that it is really *palm nut oil*. They might as well tell us "this product may contain old growth forest trees and baby orangutans", because that is what is being killed to grow the palm nut trees.

Whenever your hand goes to a food item on a store shelf, you should bear in mind the words of the Cape's favourite organic farmer (see chapter 10), Angus McIntosh: "Every time you choose a food, you are selecting a farmer, either a destructive one or a regenerative one."

One of the greenest things you can do in your kitchen – once you've started buying right – is to separate your waste. Soft, single-use plastics go into a holding bin or bucket and from there into making eco-bricks. In our house, we have taken to buying otherwise unnecessary two-litre bottles of sugar-free ginger beer or soda water to use for these eco-bricks.

You'll be amazed at how much you can stuff into a two-litre bottle; I see it as a challenge. This is not the final solution for soft plastics, but it beats having them littering the veld or blowing into the sea. (Mock us if you will, but in our home, we even wash dirty garbage before we put it into eco-bricks or the recycling bin.) Find out who in your neighbourhood is collecting eco-bricks or if there is a plastics recycling operation nearby, and drop them off with all your other sorted recycling.

The one thing you should never throw away is organic waste. It should all go into a compost bin: You can use any lidded container in which to place food scraps – including coffee grounds and even paper towelling used to wipe crockery or pots and pans clean of grease. It's good for your compost and your washing machine.

Composting could be its own discrete topic, but it's not rocket science. In fact, it is the beginning of a complete lifestyle upgrade called gardening. All soft gardening waste and organic kitchen waste goes to make wonderful compost. Some even add pet "ends" (not cat litter, though, as the anti-bacterial agents will kill off the good bugs that convert waste to humus). Just try to avoid acidic evergreen plant material.

You can buy a compost bin at any garden shop, or make your own with wood offcuts. Very briefly, it should contain 80 to 90 per cent plant matter and 10 to 20 per cent animal waste, which could be anything, even dead rats (collecting road kill as catalysts for your compost would be considered extreme, even by whole-earth crusaders). Your compost pile should be moist but never wet, and warm, as well as loose. When composting, dig in or cover new matter, so it doesn't attract flies or emit odours. Grass cuttings or leaves make a good topping.

Within a few months, summer or winter depending, microscopic workers will turn all that mush into one of nature's miracles, rich and clean "super soil" – black gold. As with recycling, just make a start, and you'll find it's easier and more rewarding than you imagined.

Now that you are producing compost, you'll need to take the next step: a garden to use it up. A garden can be as big as a farm or as small as one pot on a windowsill, and they all crave compost. Just be careful: Once you start gardening, it's like a drug, and you'll find one small tomato vine leads to heavier stuff, like carrots, beets and squashes.

The reward of eating food from your own little farm is one of nature's greatest gifts. It grows on you. Gardening will be one of the largest non-fiction sections of any bookstore, so mosey on down to yours and choose one.[3] One word of warning: When you see the eyes of your dinner party guests glazing over, it's time to move on to another topic.

And finally, we come to the malodorous subject of disposable nappies. There is no easy answer about how to deal with used ones, which is why some people simply drive down a quiet road and toss them into the veld. In many parts of South Africa, they simply go into the street in a bag, are flushed down toilets or thrown into stormwater drains. An infant will use typically about four to six a day depending, a toddler three or four, which is a lot to deal with when you pile them all up. It's also a huge headache for both the stormwater and solid waste people in our municipalities.

The first responsible thing to do is to flush the solid matter down the toilet (best when you go yourself, in order to save on water), and put them daily into a separate garbage bag. Whatever else you do, do not ever put old nappies into recycling bags or bins; the health issues down the line are ghastly.

The second responsible thing to do is to not use them. In this age of maximum convenience (never too hot, never too cold, never

---

[3] There are many good gardening books, but the best I have found is *Jane's Delicious Garden* by Jane Griffiths (Sunbird, Cape Town, 2009). Come to think of it, I used to publish the annual *South African Gardener's Almanac* back when I was a flower child.

hungry or thirsty, never inconvenienced in any way), that's a big ask. But these nappies will not go away, not for centuries to come. They're like human radioactive waste that we throw into a big hole, block our noses and leave for someone else to deal with in the future.

In the mid-1980s while I was studying all things environmental, I recall reading an article in a scientific journal about a group of architectural students at Cambridge University who, for a project, had attempted to design and build a self-sustaining home. They came pretty close, although admittedly, affordability was not one of the criteria. They found they could re-use or recycle just about everything excepting the human occupants. They had become so contaminated by the over-consumption of noxious chemicals, primarily table salt, that the fumes when burnt would have exceeded national health levels. (One wonders how they tested this!)

More recently, students did something similar in a joint project between Stellenbosch and Cape Town universities, as part of an international design competition for tertiary institutions around the world. House Mahali from South Africa was one of 18 set up in an eco-village in the desert north of Marrakesh, Morocco. The "house house" (in Swahili, *mahali* means place or house) was the only entrant from sub-Saharan Africa, and it was awarded second place.[4]

In traditional Saharan style, the 250-square-metre home faces inward onto a central, fountain-cooled courtyard – using recycled water, naturally. Incorporating the latest technology, House Mahali has a tensile-fabric roof (like those you might have seen covering some sports stadiums) that provides shade and collects water from rain and condensation. The outer cladding was made by previously unemployed women in Franschhoek – 558 wall panels hand-woven from reclaimed plastic shopping bags.

"Because the ... house was constructed with the guiding principles of circular resource usage, biomimicry and solar technologies, our house is not only cheaper [than a conventional one], with an exponentially lighter carbon footprint, but it will be able to pay

---

[4] https://www.news.uct.ac.za/article/-2019-10-01-best-little-sustainable-house-in-africa; https://www.sustainabilityinstitute.net/5587-team-mahali-rethinking-the-housing-of-africa/ (accessed 31 August 2020)

for itself through the retail of surplus energy and carbon credits," Michael Louw, team leader and senior lecturer in the UCT's School of Architecture, Planning & Geomatics, explained.[5]

For off-grid dreamers, pod housing has become a thing these days, beginning with the conversion of old shipping containers into stores (highly favoured in the townships), barber and hair-dressing salons, informal food outlets, even offices and what are called "tiny homes". Some problems include how expensive they have become, how much work and materials are required to make them comfortably habitable, and the high structural costs if they are to be mobile.

South Africa has a growing number of pod-home manufacturers, some still using the basic shipping container as the frame, while others are ground-up designs. At the top end are some Frank Lloyd Wright-inspired modular homes selling for as much as R4.5 million. Next come the mobile pods, with a lot of the cost absorbed into making them structurally and legally worthy for road transportation (which is also a design limitation).

The alternative is to build your own pod home, or complex of individual pods. Whatever you decide to do – or not to do – to help keep Mothership Earth ship-shape, think of 7.5 billion yous, and you get the magnitude of the multiplier effect. Most primary schoolers these days learn a new holy trinity of R's as opposed to the traditional reading, 'riting and 'rithmetic (although one hopes not at the expense of). The new code is recycle, re-use and reduce.

Living green is all very well, but what about all those old and worn out vehicles – our tired, diseased and toxic bodies – when they've blown their last gasket and are headed for the casket? The only thing we can say reliably about all of us is that we are all going to die. And even in death we end up cluttering the works.

Many cultures burn their dead, which uses lots of unnecessary firewood. Among India's Hindu population there used to be a custom known as sati, which required a widow to climb on top of her deceased husband's funeral pyre: thankfully, this practice has been outlawed

---

[5] https://www.news.uct.ac.za/article/-2019-10-01-best-little-sustainable-house-in-africa (accessed 31 August 2020)

for some time.[6] All the burning of wood and spouses aside, the fumes from a burning human body are truly noxious, as we learnt above.

Some more spiritually enlightened societies feed their dead directly back into the nutrient cycle, usually to vultures, but in some places hyenas. I'm guessing this means of human disposal will not be an easy sell to everyone, no matter how practical, but it does put a new spin on the conservation definition of a vulture restaurant.

But what if a tree were planted on each of the estimated 300,000 new graves each day worldwide? Instead of burying our dearly beloved departed ones inside brass-festooned wooden coffins in vast acreages of good land, we dig a hole for them, throw in a bag of compost and plant a tree?

I've thought it all through. Indigenous trees would be preferred, in order to regenerate or create natural woodlands. But other flowering species as well as any fruiting ones would be acceptable, to provide food for insects, birds and hungry people. The hole should be just 1.5 metres deep, one metre for the tree and the extra bit for the body.

The burial containers would need to be strong enough to carry the weight but ones that will biodegrade quickly, preferably a cardboard box with some light timber bracing. The unique selling point would be the name: Remember that the first person to patent this product and copyright the name is going to be the next green billionaire.

My best shots so far are the Burial Box, Bye-Bye Box and the Snuff Box, but I'm gifting them to you. All I ask is that you send a commensurate donation from each sale to a conservation cause – and that when it's my turn, I get a yellowwood.

---

[6] Although the British tried to ban it as far back as the mid-17th century, it was preserved mainly among the upper castes, and finally completely outlawed only as recently as 1987.

Going Unplugged

Goodbye, Farewell, Totsiens, Hamba Kahle and Sala Kakhule – We're Headed Down the Road and Off the Grid

DON'T YOU SOMETIMES JUST WANT to stand up, scream and chuck it all against the wall? Tell the boss they can upstick the job and the bank manager the mortgage same-wise?[1] Inform the council you're done with their rates and the tax collector that you've payed your last instalment on the gravy train?

It's really not that hard to do, but it does require a significant recalibration of who you are and what you want to make with what's left of your life. With the realisation of diminishing returns, increasingly, people are quitting the urban din and downsizing to the *platteland* as their retirement plan.

The first stratagem is to avoid those voguish dorpies, like Darling and Clarens, that have outpriced their value; even more so once-delightful places like Hermanus and Knysna, which long since traded most of their charms for the lure of development and increased rates income.

For people who like wide-open spaces, extreme temperatures and negligible rainfall, the Karoo is a big puller. Places like Cradock, Vanwyksdorp, Nieu-Bethesda, Rhodes Village and Sutherland

---

[1] A famous telegram sent by Australian humourist Lennie Lower to his publisher after an argument read simply: "upstick job arsewise stop".

each has its own appeal, as well as challenges. In the latter two it's finding enough firewood in areas virtually devoid of trees, to survive winter. In others it would be the competition of opening yet another quaint guesthouse, restaurant or art studio. Yet the gurus tell us competition is good.

In Cradock, Harrismith and Graaff-Reinet, it's a lack of water. All along the Vaal it's an overflow of sewerage. But there are so many dorps from which to choose (literally, Ariesfontein to Zuurfontein), there is no need to rush things ... unless, as was the case when Great Trek leader Piet Retief fled Grahamstown, you are skipping ahead of the law. The smart way is – and this is really the crux of downsizing to the country – sell your city pad first and then rent in the place you fancy to see if it ticks the boxes.

That is one plan, but it solves only half of the issue of getting out of the grey urban jungle. Freedom, as the old song tells us, means having nothing left to lose. To free oneself of all the bonds that tie us to the relentless machine of servitude, one has to do what Bill Mitchell from Oudrif in the Western Cape did: no boss, no rates and no credit, not even a bank account. He did eventually have to wiggle a bit around his original intents in order to run his eco lodge.

To reach his redoubt, you leave Cape Town on a reverse trip down the "telegraph road",[2] dodging sprawling townships and oil refineries, over groundswells of wheatfields and vineyards, through passes and fruit-filled valleys before you sneak around the dark orange bulwarks of the Cederberg to ascend the rolling back of the Nardouwsberg, alternatingly stony and sandy and quilted with rooibos fields. From the crest the gravel road snakes down the northern slope, following fantastical rock formations into a parched valley.

The Nama word 'nardouw' means narrow passage and, until the 1960s, when the diesel bakkie replaced equine power, it was a donkey cart route to a ford across the Doring River. Road becomes track, and you get ample chance to practise the stop-open-farm-gate-close-farm-gate-go routine. It's said some farmers can do it with their bakkie in crawler gear without ever having to stop. The ruins of old

---

[2] The Dire Straits song "Telegraph Road" tells the story of a wilderness trail that became a cart track then a highway.

farms, shaded by gnarled gum and pepper trees, give notice that this is hardscrabble country.

Your mind drifts. It's a journey informed by birds: a complaint of crows getting away with murder in the fields, a startling of pied starlings lifting off a fence, a diffusion of Cape turtle doves exploding from a copse of eucalyptus trees, an elegance of egrets gliding along the Olifants River. Once into the dry lands, lone sentinels delineate the passage: A goshawk lifts off and circles, a rock kestrel hovers, a bee-eater hops ahead of the car, a golden Cape cobra disappears into the scrub on the side of the track. It's February and a furious little dust devil kicks up its heels.

If you wanted to evade the law, the tax collector, a bank manager or any other demons of modern urban life, this would be as unassailable a redoubt as any in which to hold out. The area can be as dry as crushed bones and hot as volcanic dust in summer, then drenched and icy in winter. The Doring River demarcates the northern limit of the Cape Fold Mountains, as well as the most southerly latitude of Namaqua's spring flowering pageant.

The old ford across the river, Oudrif, has been a passage for millennia for one simple reason: when the seasonal watercourse begins to falter towards the end of the year, there is a string of spring-fed, deliciously cool natural pools stretching for about a kilometre where a sandstone ridge forms a natural obstruction. For the hunter-gatherers who left their venerated images on rock overhangs higher up on the ridge, these perennial pools would have been an important and even magical "place of power" on their eternal wanderings across the great dry beyond.

Oudrif is also the put-in point for winter white-water rafting trips when the Doring pours out from the Groot Winterhoek mountains to the southeast. And it was where Bill, a river guide for many seasons, chose to stand fast and make his happy place.

"When you start shouting at your clients, you know it's time to move on," he quips. He does not hate city life; in fact, he quite likes it, he assures me. "But what do you want from life? A great experience," he provides the answer. "Being a guide for so long, I was half bush anyway."

He'd like to retire here in the Cederberg he says, but one cannot predict the future, or stress about it. This is in reference to his 10-year lease on the farm. "Then again, the farmer would be crazy not to renew it," he says, indicating the hard landscape. "What else are you going to do here, farm scorpions?" The sardonic landlord at the eco lodge called Oudrif lifts his eyebrows.[3] The average annual rainfall on top of the Nardouwsberg is about 350 millimetres, while down in the valley it is a mere 150 millimetres. In 2019, they received a pitiful 47 millimetres.

On first meeting you could take Bill for a bit of a bumpkin, in old (but clean) T-shirt, shorts and sandals, with a pirate earring and flat South African accent. However, if you want to see what's really inside a person's head, check what they read: On the bookshelves in the lounge-dining area you'll meet a cornucopia of characters – from birds and plants to economists, politicians and detectives. Conversations flow easily, sometimes droll, sometimes witty, but always well informed.

The wisdom of Bill's seemingly content life seems to be to not over-complicate things. This natural metaphysician keeps his conversations short, to the point of abruptness. It might just be the 20 years of bush living (following a degree in industrial sociology and years odd-jobbing around the world). Then he found the place that resonated with him.

With his best mate, Paddy, they jumped in. "We didn't talk about it much, just went ahead and did it really," recalls my twinkle-eyed host. "He was the capital, and I was the labour." From the start, they wanted to be off the grid and sustainable, "Call it green, I suppose."

It had to be as environmentally sustainable as possible but, since there is only sand and no clay anywhere near, cobbing (building with a straw-clay mix instead of traditional bricks and cement) or anything like it was a non-starter. But it also had to be as inexpensive as possible – they had capital but not wads of it. Plaster and floors had to be cement rather than clay: "Not as eco-friendly as you'd want," Bill half smiles, "but a man makes do." Followed by, "I was no kind of handyman, in fact I'd not built anything previously."

---

[3] https://oudrif.co.za/ (accessed 31 August 2020)

The partners chanced on a newspaper article about Greenhaus – led by alternative architect-builder Etienne Bruwer, champion of straw-bale building – and realised they'd found their man and their plan. They sited the lodge on a natural but overgrazed terrace above the river, siting each stand-alone unit as much as they could on rock.

They did not want to – and indeed did not need to – do any earthmoving, with all the work being done by hand, "unfortunately", Bill jests in his dry way.

The straw came from a nearby wheat farm; the structural timber was bought at R5 a metre by clearing gum from a dry streambed (by hand, of course). Doors and windows came mainly from demolitions, the classifieds and regular visits to the treasure trove of just about every owner-builder in the region, Ross Demolition.

"We learnt very quickly how to sort the crap. We sorted through a lot of crap."

With straw bales, you first assemble a timber frame, and then fill in the empty spaces. You can use a chainsaw afterwards to cut holes wherever you want them. "There was straw everywhere, in our clothes, in our hair, in our mouths."

Since there were no foundations, they had to start by embedding metal-spike footings for the bales. They started using rebar but found out almost immediately why it's good for keeping skyscrapers upright; fence dropper posts soon became their preferred footings. Then all the straw had to be compressed (using car jacks and fence wire pullers), covered with chicken wire and then sewn together with baling twine, then finally plastered. "The chicken wire was our biggest expense."

The roof is also made entirely from straw on a timber frame, and then plastered. Here also they had to settle for a cement capping. "Not the greenest solution, but I'm glad we did, or I'd be a full-time repair man." At any rate, May each year is still roof maintenance time. Due to the extreme climate, the cement plaster cracks and leaks.

"I don't mind so much, but some of the guests seem to," Bill says, deadpan.

"We planned the place for 10 guests. That number of people can disappear around here, but not 20. We wanted peace and quiet for our guests but also for ourselves." The other "self" is Jeanine, partner and wife for the past 17 or 18 years. (She's not quite sure…)

Bill was certainly not looking for a wife when he went to drop off a load of washing at a farm on the lush side of the mountains, and neither was Jeanine in need of a man. Her hands were more than full at the time, working as a shepherd and bottle-feeding 80 kids through a hard winter, and then rescuing wild creatures. She'd bolted from the big city ahead of her own pack of hellhounds.

When he went to collect his laundry, he invited her to Oudrif for the weekend, and she never left. In order to fund her work with EnviroVet community veterinary clinic, Jeanine sells beauty products that are home made from locally sourced herbs. (You'll also find these products in the guest rooms.) Her feminine touches are all around. You can see her presence in Oudrif's food garden (well fenced against marauders), the food, and the many animals that come and go.

There was the baby thick-knee named Elvis, because one leg never stopped shaking. There was Bo-sheep that chased guests, raided the kitchen, butted into dinners and stripped toilet rolls, driving everyone crazy. She had to go. Then the two spotted eagle-owls Beryl and Boris. Beryl would swoop in on silent wings and steal food off dinner plates – "what the…!" There was Dave the steenbokkie, there were mice, and there was always (and still is) a mob of dogs.

We need nature to make us truly humans, she tells me as we dodge the smoke around the braai. We need to understand how we fit into the bigger pattern. "To use less, so we can live more. To be able to do that in the bush," she says, gesturing around, "that's magic."

Magic: when the afternoon mercury passes 40, a dip into the cool wedge of mountain water seems inexplicable, sybaritic, in this austere environment. It too is a place of birds: reed warblers thrilling in the dense phragmites reeds, swallows skimming the surface and doing aerial acrobatics to hawk insects on the loft of a breeze that ripples the water and flutes the reeds.

You can laze there until the redolence of a braai comes wafting down from the communal building. Every evening, when guests are in house, there are lamb chops and boerewors. Despite being a vegetarian, Jeanine cooks the meat – it's part of the job and the bush experience. It comes with fire-baked bread and enough salads and vegetable dishes not to miss meat anyway.

As the moon rises over the hills and plates, the pool's pewter surface with copper, it seeps into your being that this is certainly a place that could soothe a weary soul. But I'm working here and have to drag myself back to the off-grid nitty-gritty.

The guest units have electric lights and gas geysers, and the entire lodge runs on a solar system. Best thing about that, I am told, is how much more efficient and cost-effective solar has become over the past two decades, from around R45 a watt to R8. Water comes from the pools and sewerage goes from bio-enzyme-treated septic tanks to French drains. Bill reckons it would take around about 20 years for it to seep through the sandstone bedrock back into the river. "That means our first flush should be getting there about now," he says.

"I'd love to be off the internet. It sucks up time, but we need to get people to come and stay here," Bill says, as a parting observation. Later, when I send them a mail to check some information, I am tickled by their auto response message: "Thank you for your enquiry. Please note that we are not office bound and completely off-grid … communication systems are not always reliable. We appreciate your patience. Have a beautiful day."

THE PATH TAKEN BY ARTIST-CRAFTSMAN Gerick Terblanche and his partner, jeweller Jana Gabelmann, was more one of need than the want that motivated the road to Oudrif. The couple met and made fast in the clubs of Cape Town, he a struggling artist and she a visiting German on an internship in adult basic education and training.

Things were going really well, until the day their apartment was burgled. The urban rat race, compounded by memories of an earlier mugging (of Gerick) at knife- and gunpoint … Increasingly, their eyes began to see only the darker underbelly of the city: The poverty, the filth, street crime, the noise, the traffic, the congestion seemed to pile up into an insurmountable obstacle to personal safety, health and happiness.

They wanted to stick together, but not there. (The story of their courtship, recounted with candlelight and wine, was a long and winding interlude.) Some years previously, Gerick's folks had bought a portion of the farm Snyderskloof, off a road that branches off the route between Matjiesfontein and Sutherland, which is to say pretty much the end of the road, on a ridge-line somewhere between the Tankwa and the High Karoo, or Nuweveldberg.

On the piece of weekend bolthole land, there was a simple bywoner (tenant farmer or foreman) cottage and a shearing shed, the latter converted into a Karoo-style farmhouse some years earlier by Gerick's parents: You'd never guess it had not been built, and furnished, by their great-grandparents. Gerick's involvement with the farm inspired him to move out of the city and build his own place on the land that was available there.

For Jana, with no close friends or family in South Africa, the move out of town was an easy one to make. "She was living in a dump in Obs without any understanding of the local rules, mainly when it came to issues of personal safety," Gerick cuts in, although patently Jana did not see it exactly that way. "And no more having to drive," she adds. They agree that leaving town was a case of leaving stress behind and finding space and – more importantly – freedom.

"Here we are together pretty much 24/7. It's a huge commitment, but I reckon we are doing pretty well," the artist offers. The jeweller concurs. "We respect each other's space," she says. Gerick is building her an autonomous studio pod, so she can move out of the cluttered cottage-cum-storage room that is serving as a cubbyhole studio. He initially planned to use re-treaded clay bricks for the studio, but later settled on rammed earth and Cape reed instead.

Another commitment is living off the grid, Gerick admits. Off-the-grid living is what made my path cross with his in the first place. Some people would call it coincidence, others synchronicity or perhaps synchro-destiny, but it was one of those things. My partner and I were making our escape from a soirée, squeezing our way out the entrance, where we introduced ourselves to another couple also making their getaway.

"You?"

"Artist, and you?"

"Writer."

"What?"

"A book on environmental issues, about people who have gone off the grid, but I've found only one so far."

"Hey, you have to visit us, but not right now. We should be ready in about a month."

I arrive four months later, in February, which is a good time to get an idea of the challenges out there, primarily heat. "Winters here are also challenging, especially when you're living in a pod," says my host. Another, but mainly for his collaborator in alternative living, is a new studio and a new home, one half-built, the other still only at foundation stage.

"Another month," says the very optimistic and ebullient artist-builder. The collaborator smiles benignly. "He is a perfectionist," I am told. "Which can be a synonym for procrastinator," I suggest. Two heads nod. We stroll over to the building site, which is one of several pods planned for the operating domain of two creative people.

Each will be built using different materials, just for the fun and experience of it. I'm a wannabe green-house builder myself, so we talk shop: battery input and output, the pros and cons of air-conditioning in the Karoo (without it, there's no working indoors with a blowtorch during summer), recycled insulation, sourcing used timber.

"There's no excuse not to go solar now," Gerick says. "You just have to use it wisely. You don't switch on every appliance at the same time. And you use less." For starters, you can cook on gas, or wood if you have a sustainable source. "And you don't need a 'V-8' stove for just two people.

"The big question is, do you care?" Gerick asks. We're standing in the shade of the half-completed building and staring out across the seared Karoo landscape. "If you do, then it is not hard to convert to a conscious and caring lifestyle. In towns people get addicted to gadgets, to stuff."

One huge find (sustainable builders are forever on the lookout for materials they can repurpose, like newlyweds shopping for furniture) was a consignment of seamless tempered glass panels left over from a high-rise job in Cape Town. When you get that kind of windfall, you don't say, "no, it doesn't fit the design" – you redesign. Luckily, with a timber and board building, you can just take a saw and cut new holes where you want them.

We pick our way over the gulch to what is currently the building materials storeroom, where again we stand in the shade and chew the chaff. The place is destined to be a workshop, I am informed.

"I don't think of myself as being only an artist," Gerick declares. "I want to make things like furniture, knives…" Artist he certainly is, but his ambition to be a craftsman explains his super-realistic animal and bird paintings with various abstract but finely worked backgrounds that represent whittling (chiselling walking sticks and utensils, traditionally with a pocketknife), another of Gerick's past-times.

From there we brave the afternoon heat to make a dash across to Jana's workspace. And there she is with blowtorch, crafting her unique Karoo creations inspired by light, space, rain and rainbows. She walks the veld fossicking for bits that she can incorporate into her silver and brass pieces. She sells them online, but also makes regular trips into town to resupply the shops that carry her Loved by Elli (her mother back in Berlin) range.

When dinnertime comes, the whittler is also a pretty deft cook: venison steaks on the braai along with crisped potato slices, polenta cakes with a bean, tomato and onion ragout, and home-baked bread. He shot the blesbok himself, with bow and arrow. "It's all part of living off the land," he tells me.

He dreams of opening a restaurant on the hill way above the yard. "Not really a restaurant so much as a culinary experience, a conversation about food and environment, including the role of hunting, and the Karoo lifestyle. And spending two hours cooking is far nicer than spending it commuting," he muses.

This is the Karoo, so after dinner is talking time, with the fire crackling, when you share stories and spin philosophies. It's just that kind of place – vast, spherical, with the overarching Milky Way resplendent on its dark cloche. It's the reason one of the world's pre-eminent astronomical observatories is sited "just up the road".

Gerick recalls getting going on the building, hiring itinerant workers who were the living incarnations of the people who used to follow the rains and the game across these reticent plains. The theme was about how eerie and folkloric the place can be. He was about to start working on the old shed and needed a handlanger (assistant). On cue arrived Boetie Baadjies, festooned with crude tattoos.

The ink work looked dodgy, so Gerick asked if he'd got them in prison. Yes, the wizened Khoi man confirmed. He'd been imprisoned for murder. Three times. One night, actually it was early morning, there was a huge commotion in the shed where Boetie and the other labourers were staying. They heard shouting: "Los hom! Los hom!" (Leave him! Leave him!") Chairs were flying and there was a general melee. Summoning all the courage he could, the contractor approached the workers' shed.

Gerick waited for the chaos to settle and then went in and asked what was up. "The ghost came," the hired men told him. "She is a very strong woman." The men all wanted to pack up and leave right then.

"The Karoo can play tricks on your mind," I am informed, maybe warned. "You cannot judge the local people for what they believe. This is their world, and it is all very real. You don't make jokes about it here."

We sit in silence for some time. "The beauty of being isolated is that your creativity is totally independent of outside influences. True originality can only come from real isolation."

Silence. "That is what living off the grid really means."

Maybe this is the centre of the universe, I offer as my best shot. Considering the outer spiral arm of our galaxy, it sure feels like it. The wine is finished, and it's time for me to go dream about my own escape from the rat race.

ON MY PEREGRINATIONS AROUND THE country over many decades I have always been on the lookout for a perfect bolthole. At various times I have fancied, among many, Morgans Bay, Cape St Francis (the wild side, not the diamonds-and-villas marina side), Cintsa, a fynbos plot at Pringle Bay (I actually bought one, but life got in the way and I had to sell), Storms River, a Little Karoo game estate called Touwsberg and, more recently, an eco estate in the foothills of the Baviaanskloof Mountains called Honeyville.

My problem seems to be that, every time I am about to commit to my own Elysian Field, I move the goalposts and my dream off-the-grid future skips on down the road. My partner regularly brings me back to reality, pointing out that we live on a perfectly peaceful waterway in easy reach of a beach, a mountain and a forest. Which is the truth. But my response (if only to myself) is, if we all thought that way we'd never get our people to Mars.

If Men Could Fly
What I Could Not Say

WRITING IN THE MID-1800S, AMERICAN contrarian and naturalist Henry David Thoreau[1] observed that, if humans could fly, they would likely lay waste to the heavens much as they had done to Earth (brace yourselves Martians).

There were so many things I had wanted to write about in this book but could not: There was no shortage of environmental issues to cover. But sometimes, you have to spare your audience and say that's it, for Pete's sake, enough, the end. And anyway, I had reached the publisher's word limit.

As with my other "stories from the veld", I wanted to cover all the obvious subjects but in as unexpected a way as I could. Moreover, I wanted to cover the unexpected, like the chapters on lies and how they have impacted our outer as well as our inner lives.

When beginning the project, I asked my peers what subjects they thought should be covered. One that came up frequently – but that I was unable to fulfil – was the rape of our estuaries by illegal fishers. You can buy a small-gauge fishing net at just about any Chinese shop. You can buy just about anything at most Chinese shops. They are the tips of the tentacles of the emergent Chinese empire.

---

[1] Henry David Thoreau (1817–1862) is the author of Walden; or, Life in the Woods. It is one of the most influential books in the entire canon of nature writing, but not everyone gets it. It is contradictory, often confusing and obscure, but always inspired. He greatly influenced the environmental movement, but also luminaries including Leo Tolstoy, Mahatma Gandhi and Martin Luther King through his writings on civil disobedience in the face of immoral laws, especially slavery.

I live on an estuary, the water of which laps our front garden. I am not a fisher myself (my mate Donald reckons it is no better than snaring animals for bush meat), but there is a gang of them, honest ones, who patrol our stretch of coastal water at night looking out for illegal netters. They can confront them, but they have no power to apprehend them. So, the netters simply move off and come back another time. But at least we have eyes out there.

Another man who shoulders a giant's load of the work policing our coastal areas, in this case in the Eastern Cape, is conservation officer Div de Villiers. He is one of the people without whom you worry about what the state of things might become. Whether it's illegal fishing, chopping down precious hardwoods in the coastal forests, trapping animals for meat and muti, or illegal sand mining, De Villiers is one of the brave "green scorpions" on the ground, and also sometimes in the air.

I was on my way to visit him when we were hit by COVID lockdown and so – extremely reluctantly – decided I could not do justice to his story without going through the rites of passage of sharing a braai, a cold drink and joining him on a recce trip or patrol.

Another person I was set to meet on that trip was Bruce Mann, a senior marine scientist at the Oceanographic Research Institute in Durban. By phone and emails he told me how he had tried to introduce and manage a community fishing project at Lake St Lucia, South Africa's largest estuary, which ended in a fiasco.

As soon as fishing restrictions were lifted, the communities living around its shores waded in. Some did so according to the rules, but others did so according to the intrinsic principle known as the tragedy of the commons: I must take as much as I can before someone else does. This is the same natural law that leads to the pillaging of our oceans, forests and shore zones.

The intertidal rock beaches of the Eastern Cape are probably the hardest hit of all the ecosystems in South Africa. Years ago, places like Hamburg and Dwesa had some of the most diverse, enrapturing rock pools of our entire coastline. If you knew them back then, going there now could break your heart. Someone I know visited Hamburg recently and reported the shore was bare, denuded, stripped, shorn, shaved of every living thing. There was just bare rock.

There are laws, but there are also poor communities living along that coast for whom the sea is, and has always been, a free source of food. How can we say to them: Yes, we hauled out all the big fish for holiday fun, and we stripped the bays of chokka for our restaurants and for paella in Spain, but now you mustn't take the mussels and the oysters, the crayfish and the octopuses and even the whelks and limpets, to eat?

What is happening to the rocky shores of the Eastern Cape is similar to the perlemoen poaching that has virtually wiped out abalone in the southwestern Cape. It won't stop until there is nothing left. In truth, with marine molluscs and crustaceans, they are unlikely to go extinct because there will invariably be stock out of easy reach that will continue to spawn. But the populations will crash, and then the socioeconomic conditions that led to the collapse will also implode, and then what?

In the case of the abalone, it will have an impact on the drug gangs of Cape Town that, via the perlemoen poachers, procure drugs from the Chinese traders and fishing boats,[2] who export the perlemoen to Hong Kong where they are considered a delicacy. And when the drugs run out, and all the money that came with it, what are gangs going to get up to?

ANOTHER TOPIC I HAD HOPED to cover was the controversial and community-splintering case of sand mining at a place called Xolobeni on the northern Wild Coast.[3] To get there, you turn off the N2 towards Mkambathi and Mpetshwa, then down a succession of narrowing gravel roads and tracks that negotiate the hilly, tortuous terrain that lies pressed between the Mtentu and Mzamba rivers. The final approach to the village, comprising several traditional Xhosa

---

[2] From my reading of the book *Poacher: Confessions from the Abalone Underworld* by Kimon de Greef and Shuhood Abader (NB Publishers, Cape Town, 2018).

[3] Information for the section on Xolobeni was drawn from https://en.wikipedia.org/wiki/Xolobeni_mine (accessed 20 September 2020) and https://mg.co.za/article/2016-02-12-we-will-die-for-our-land-say-angry-xolobeni-villagers-as-dune-mining-looms-1/ (accessed 20 September 2020).

(or, more correctly, Mpondo) kraals or iziqhulo, is along little more than a cattle track.

The fossil dune sands here are brimming with heavy minerals. The local subsidiary of an Australian-based mining company (Transworld Energy and Mineral Resources), wants to dig up the dunes, wash the sand to reclaim an estimated 9 million tonnes of ilmenite and titanium-iron oxide, as well as lesser amounts of rutile, zircon and leucoxene, which are used mainly in making paint and paper white.[4]

The problem is that many of the people around Xolobeni don't want them to. They want to be left alone to graze their cattle where their ancestors did and, if there is to be development, it should be of the ecotourism kind. There are, however, people further up the local food chain who very much want the mine. For their liking, ecotourism is not corruptible enough.

A lot of money has already changed hands in the area to help pave the road, literally: A major upgrade of the N2, including two massive bridges, which just happens to bring the national road close to still-isolated Xolobeni, and which is believed to be part of the bigger conspiracy. Regardless of whether or not that is true, there has been a lot of skulduggery, including intimidation, violent attacks and even murder of the people standing in the way of the mining interests.

"The people" even took the Minister of Mineral Resources Gwede Mantashe to court, where they won the right to be properly heard and consulted (already the mine and its lackeys have mangled the implications of "properly") and to give their fair democratic consent to any kind of extraction on their traditional lands. But they are split in two, and that partition is being leveraged to divide and rule.

You know you are up against hard odds when the person you took to court is the minister of mineral resources and mining. This sad saga has been going on for more than a decade and there is no end in sight.

But Xolobeni is only one of many land issues being fought across the land. If you thought the government's motivation behind the controversial plan to implement land expropriation without

---

[4] /en.wikipedia.org/wiki/Xolobeni_mine

compensation was a ruse to get their hands on white-owned farms, you are politically ill-informed: It's about the much more extensive tribal lands and who controls them.

One-third of all the land in KwaZulu-Natal is in effect owned by Zulu King Goodwill Zwelithini, through a set-up called the Ingonyama Trust. Not only is the monarch paid a king's ransom as a salary from taxpayers' coffers, he also extracts rentals from the people who live on that land.[5]

But now some of the people who live there are getting uppity and doing what they did in the Monty Python film *In Search of The Holy Grail*. They are saying: "You aren't our king, we didn't vote for you," and also "You cheated us when you got us to sign new lease agreements that basically forced us into strangulatory lease agreements or be kicked off the land of our ancestors."

Just how we came to have a Zulu king in a modern socialist democracy of sorts is odd indeed, but he is one of only six monarchs, no doubt chosen by God but paid by us, the hard-pressed workers.[6] King Zwelithini alone receives a salary approaching R70 million (and every year moans that it's not enough), on top of more than R100 million a year from land rentals.[7] There's a story there for sure, but maybe for another time.

And speaking of human effluent, I forewent a chapter provisionally

---

[5] https://www.dailymaverick.co.za/article/2019-12-04-death-threats-and-fear-rule-as-scramble-for-pristine-tribal-land-continues/ (accessed 20 September 2020) and https://www.dailymaverick.co.za/article/2019-11-20-eleventh-hour-delay-in-ingonyama-trust-court-case-involving-king-goodwill-zwelithini/ (accessed 20 September 2020)

[6] During the internecine battles waged between pro- and anti-ANC factions in the lead-up to the democratic elections in 1994, the ANC sought support from "traditional leaders" to help counter opposing forces, most notably the Inkatha Freedom Party in KwaZulu-Natal. That, in a nutshell, is why we now have a Traditional Leaders parliamentary act that legitimises our tribal kings and chiefs. However, now that many common people on tribal lands are vexing under burdensome taxes and rates foist upon them by these leaders, in my view the ANC is having to tap-dance around its socialist principles and realpolitik, as well as carefully counting the votes on each side of that factional fence.

[7] https://www.dailymaverick.co.za/article/2019-12-04-death-threats-and-fear-rule-as-scramble-for-pristine-tribal-land-continues/ (accessed 11 September 2020)

titled "Where Does All Our Shit Go?" Once you have finished with your daily purification and flushed the toilet, that is only the beginning of a long journey undertaken by your bodily wastes. You probably never think of it again, but other people have to.

From your loo, it disappears underground into a vast network of pipes of ever-increasing size that link every flushable toilet in the city or town. Along the way, the reticulation also captures all the bathroom waste, most of our kitchen wastewater and much of the stormwater along with rags, tampons, nappies, household waste (why use a dustbin?), bits of wood, other building materials including bricks, car parts, shopping trolleys, old engine oil… just about anything you can imagine that fits.

All of that, where it does not block the works, ends up in one or other water treatment plant. There it is filtered for solid matter and surface scum (mainly oil and kitchen grease) and then treated with bacteria and pumped with oxygen to help break it down further before it is sent on to some or other watercourse or directly into the sea.

That is the plan in any case. In South Africa, even in supposedly more or less functioning cities like Cape Town, the authorities seem to have lost the will to cope with sewerage, and much of it gets pumped, raw, into our rivers and oceans. That was worthy of a longer story, but you get the picture, and the whiff of it. (Some of this is covered in the chapter 12.)

It was the development of clean water supply and water-borne sewerage that allowed cities to develop into the huge, high-rise, high-density and outwardly clean entities they have become. It all happened when we still thought water was a free and limitless resource. Now that we know better, the realisation comes that the future has to be one with waterless toilets. That's all in the future, but the future starts tomorrow.

And so on to the sea. It's another thing we all thought was infinite, the lie to which was given by the industrial scale at which it has been harvested – if that is not too euphemistic a word to describe the pillaging of marine resources over the past 100 years. The custodians of our own seas have done a miserable job looking after them. First, they rescinded the licences of traditional fishing communities all around the coast in the name of job creation – mostly for their own families and friends.

Then, trying to some degree to backtrack, they introduced ill-conceived new fishing industries like demersal (living near the bottom) shark fishing that is causing the collapse of our shark populations while shipping the meat off to Australia and the fins to China. We are talking about 300 tonnes of shark fillets and fins a year, which represents tens of thousands of sharks being killed.[8]

These days, a much bigger threat lies on the edge of our territorial waters – an ever-present armada of foreign, mainly Chinese, fishing trawlers. During the day they comply with international law and have their tracking systems switched on. But come dark, off go the tracking systems, and in they come with their devastating trawling nets that strip the ocean bare to seabed level and leave a desert where before there were thriving marine ecosystems. You'd think the South African Navy, with all its new corvettes and submarines, could protect our resources from these poachers, but they don't seem to see that as being part of their job.[9]

Exactly what their job is, given that we are not expecting an attack from the Southern Ocean any time soon, is unclear. When quizzed by a foreign journalist, a Navy spokesperson thought the submarines could be used to protect bathers and surfers from sharks. Maybe the great whites in False Bay read about it, and that is why they have been so scarce in the bay these past few years.

---

[8] From the paper "South Africa's Demersal Shark Meat Harvest" by Charlene Da Silva and Markus Bürgener, published in *Traffic Bulletin* in 2007 https://wwfeu.awsassets.panda.org/downloads/shark_paper_1.pdf (accessed 31 August 2020)

[9] www.news24.com/news24/mynews24/of-subs-and-sharks-and-stolen-jets-20160302

Postscript
A Handful of Stones

I'M SURE YOU, PROBABLY MANY people, maybe even everyone, wake up some days and feel just like Dwayne Hoover. I know I do. He is the main character in a novel by Kurt Vonnegut,[1] whose books have helped humour me through decades of watching our planet burn. Hoover is a successful Pontiac salesman in Midland City, Michigan, who – because of some chemical he did, or possibly did not, take – is going mad.

The other main character in the book is Kilgore Trout, a prolific but completely unknown science-fiction writer. Thousands of his stories are published in porn – or "wide-open beaver" – magazines. One day, Hoover reads a story by Trout that informs him he is the only real person on Earth; everyone else is a robot. It's part of an experiment being conducted by God, to see how he responds. What with Hoover losing his marbles, things start to go stir crazy in the irreverent and slapstick way readers of Vonnegut enjoy.

I feel like I'm Dwayne Hoover just about every day, and have done so ever since I became seriously worried about the natural environment and what people were doing to it: How could actual biological humans do this!?[2] It seems so long ago now I cannot recall exactly when it started. I know I was still young enough to have the naivety and the energy to admonish people who tossed a cigarette butt or sweet wrapper: My family accused me of trying to change the world one litterer at a time. That seems so quaint now. Who cares

---

[1] *Breakfast of Champions* (Panther/Paladin-Grafton, London, 1974)
[2] As a fun aside, the combined use of an exclamation mark and a question mark is called an interrobang.

about litter when you can't breathe?

Now, when I have used up three score of my allotted days, I feel an urgency to find a place where I can go to ground, escape the robots, grow stuff and think. The worst of it is that I grew up as a famous vegetable dodger. I've got much better since, but things like soggy aubergine and cabbage still scare me.

I asked a mate living the good green life in Howick about what was afoot in his vale of the Midlands and learnt of someone named Nikki Brighton, who reckons she is a hard-core forager. In my (albeit limited) experience, that stuff is usually astringent, tough, furry or spiky (which is why we tamed the tastiest, juiciest ones). Nevertheless, for the greater good, I am willing to give it a go.

Anyway, the brave woman rises early each morning and walks the local riverbanks with her dog, Bean, searching for "weeds" – nettle, dandelion, chickweed and plantain, even blackjacks. My fear is, while I reckon I could recognise a nettle or a dandelion if I fell face-first into one, I would have no idea if I were chewing chickweed or henbane.[3]

I enjoy spekboom and clover in a salad or on a sandwich as much as the next goat, but I do not know if a steady diet of dandelion and nettle leaves would succour me. Just the other night I watched a documentary about Christopher McCandless,[4] the young man who famously went walkabout in Alaska trying to live off the land, only to die of starvation. And yet it is a quest I am thinking of more and more these bleak days.

My Christopher McCandless moment came when I read a book about another nutter who decided to go not only off the grid but pretty much off the map. Written by one-time economics lecturer Mark Boyle, the book documents a financial experiment: to spend a year neither earning nor spending money.[5]

That malarkey went so well, Irishman Boyle took it to the next level:

---

[3] It is thought Viking raiders ingested henbane, the "witches' drug" otherwise known as nightshade and potentially fatal, to whip themselves into bloody frenzies before going into battle.
[4] *Return to the Wild: The Chris McCandless Story* (2014) directed by Jeanmarie Condon and Ann Johnson Prum.
[5] *The Moneyless Man,* and *The Way Home: Tales From a Life Without Technology* by Mark Boyle (Oneworld Publications, London, 2010 and 2019).

Postscript

Go somewhere off-grid and live without technology of any kind for a year or however long it took. (We'll never know how long he stuck it out unless he writes another book about it.) The west coast of Ireland can be a brutal place to be without modern basics. His girlfriend eventually left him for want of a hot bath.

Still, I confess, those books gave me a brain worm. Just about every other night I dream about building a sustainable home, growing food, tending small animals, keeping bees and foraging. Boyle mostly foraged, grew some potatoes, collected road kill and poached trout from the local streams. I reckoned in sunny South Africa you could fare a lot better with a lot less struggle.

So far in my dreams I have built a stone house in the desert, cob and straw-bale pods on one or other smallholding, shacks from recycled timber in the Tsitsikamma area and even a cluster of wattle-and-daub huts on the Eastern Cape coast.

Doing it this way means I can build a different place anytime, anywhere of my choosing, and tinker with the details until sleep steals me. Like how to make mortice-and-tenon joints, install solar panels, construct greenhouses, design the labels for my honey, and ride my mountain bike down winding tracks, greeting the rural folk and beasties as I go.

I was looking through some old photos recently and found one I had taken along Namibia's Kavango River, around the Popa rapids. The photo is of a hubcap nailed to a tree, and on it is painted in crude white lettering: "Why is the time here so very fast?"

I realised then that what my brain worm was trying to tell me is that I, we, all of us, need to recalibrate time. That unknown span of time we have left. Those of us who have achieved some level of financial stability tend to sink into comfort zones where very little changes day after day.

We get up, get dressed, do whatever job we do, settle into the comfort zone at home, put on the air-conditioning, or the heating, or the TV. Same same every day, and we insulate ourselves from the vagaries of nature. It all becomes a sameness, which causes moments to pass at numbing speed and years at ever accelerating speed.

I'm thinking of decamping to some place where the years pass more slowly by the very act of watching things more closely: What is sprouting, budding or flowering? What is breeding, nesting or hatching? Seeing which way, and at what strength, the wind is blowing, watching the sun set and the moon rise, noting each fine declination of time. Something is telling me *that* is the correct way to spend my last stones.

Explorer-adventurer Kingsley Holgate uses an analogy: Sit somewhere, preferably on a beach, but a mountainside will do as well, and pick up seven stones: each pebble represents one decade of your useful life. Throw away one stone for each decade you have already spent. What are you left holding? Best you use it well.

Sometimes I feel the old joke, that of all the things that I've lost I miss my mind the most, is not so funny. One last thing I feel a strong need to regain, along with my mind, are the internal conversations that were my constant companions during the years of hiking, mostly alone, to produce several of my early books.

It is hard to heed your inner voice when you are surrounded by the clamour of a city. Quietude helps to slow down time and make the moments more momentous. I've got two stones left, and one is already well rubbed.

Thank you for joining me on this bumpy but inescapable green detour. I hope the remainder of your life's journey goes swimmingly well.

About the Author

DAVID CANNOT RECALL EXACTLY WHEN he became ozone friendly. What he does recall is worrying about things like the energy efficiency of boiling water long before he learnt anything about Watts and Amps.

School protests in 1976 lit a spark and the next year saw him studying how to be a muckraking journalist. There the author started spending weekends rock climbing, with people who seemed to know an awful lot about stuff like plants and rocks and ecosystems. This so impressed him that he went on to study earth sciences.

More than four decades and some 20 books later, he has been fortunate to realise a long-time dream – becoming a paperback writer. *Big Pharma, Dirty Lies, Busy Bees and Eco Activists* is his fourth "non-fiction narrative" published by Jacana Media and, he says, he's not done yet.

We thank the following for their support in publishing this book:

Arthur Goldstuck
Ashwin Moyene
Ben Williams
Beverley Naidoo
Carolyn Raphaely
Catriona Jarvis
Corinne Rosmarin
Denis Hirson
Dianne Stewart
Gill Bolton
Glen Impey
Graeme Friedman
Helen Douglas
James Bissett
Karin Pampallis

Kevin Ritchie & Associates
Louis Gaigher
Maeve King
Mamma Jacqui
Mary Burton
Michelina Giacovazzi
Moira Levy
Roger Southall
Rona V van Niekerk
Ryan Childs
Sebastian Seedorf
Steven Dubin
Sue Grant-Marshall
Trisha Cornelius